Electroencephalography - From Basic Research to Clinical Applications

Edited by Hideki Nakano

Published in London, United Kingdom

IntechOpen

Supporting open minds since 2005

Electroencephalography - From Basic Research to Clinical Applications
http://dx.doi.org/10.5772/intechopen.87425
Edited by Hideki Nakano

Contributors
Boulenouar Mesraoua, Musab Ali, Rola HM Khodair, Yazan Nofal, Dirk Deleu, Hassan Al Hail, Gayane Melykian, Osama Elalamy, Covanis Athanasios, Jesús Pastor, Lorena Vega-Zelaya, Elena Martin Abad, Raafat Hammad Seroor Jadah, Julián González, Almudena González, Manuel Santapau, Hideki Nakano, Liljana Bozinovska, Adrjan Bozinovski, Subha D. Puthankattil, Alexandre Magno da Rocha Vianna, Ali Akbar Asadi-Pooya

First published in London, United Kingdom, 2021 by IntechOpen
IntechOpen is the global imprint of INTECHOPEN LIMITED, registered in England and Wales, registration number: 11086078, 5 Princes Gate Court, London, SW7 2QJ, United Kingdom
Printed in Croatia

British Library Cataloguing-in-Publication Data
A catalogue record for this book is available from the British Library

Additional hard and PDF copies can be obtained from orders@intechopen.com

Electroencephalography - From Basic Research to Clinical Applications
Edited by Hideki Nakano
p. cm.
Print ISBN 978-1-83968-288-9
Online ISBN 978-1-83968-289-6
eBook (PDF) ISBN 978-1-83968-290-2

For EU product safety concerns:
IN TECH d.o.o., Prolaz Marije Krucifikse Kozulić 3, 51000 Rijeka, Croatia, info@intechopen.com or visit our website at intechopen.com.

Meet the editor

Dr. Hideki Nakano is a physical therapist and associate professor at the Neurorehabilitation Laboratory, Graduate School of Health Sciences, Kyoto Tachibana University, Japan. He received his Ph.D. in Health Science from Kio University, Japan, and has accepted research training experience as a JSPS Research Fellow at Neurorehabilitation Research Center, Kio University, Japan, and Queensland Brain Institute, University of Queensland, Australia. He specializes in neuroscience, neurophysiology, and rehabilitation science and conducts research using non-invasive brain function measurement and brain stimulation methods such as electroencephalography, transcranial magnetic stimulation, and transcranial electrical stimulation. His current main research interest is developing rehabilitation evaluation and treatment methods using visualization and manipulation methods of brain functions for healthy, elderly, and diseased patients.

Contents

Preface

Electroencephalography (EEG) was discovered in 1875 when Richard Caton of England recorded electrical impulses from the brains of rabbits, dogs, and monkeys. He reported that EEG fluctuated steadily and that these fluctuations changed with light stimulation and exercise, as well as with sleep and death. Later, in 1924, Hans Berger of Germany recorded the potential fluctuations of the human brain, and in 1929, the first paper on human EEG, "Über das Elektrenkephalogramm des Menschen," was published.

Current non-invasive brain function testing methods can be broadly divided into two categories: electrophysiological testing methods and brain function imaging methods based on hemodynamic principles. The former includes EEG, magnetoencephalography (MEG), and transcranial magnetic stimulation (TMS), and the latter includes functional magnetic resonance imaging (fMRI), positron emission tomography (PET), single-photon emission computed tomography (SPECT), and near-infrared spectroscopy (NIRS). Today, these methods are used in human neuroscience research and clinical practice, and EEG is one of the oldest non-invasive methods for testing brain function.

EEG is used in a wide range of fields, from basic research on the human brain to clinical diagnosis of epilepsy and psychiatric disorders. Recently, EEG has also been used as a tool for rehabilitation evaluation and treatment, and research and development of brain–machine interface have been actively conducted. This book presents basic research and clinical applications related to the aforementioned topics and is organized into the following eight chapters written by experts from around the world.

Chapter 1 "EEG-Emulated Control Circuits for Brain-Machine Interface"

Chapter 2 "EEG Analysis during Music Perception"

Chapter 3 "Multicriteria Algorithm for Multisensory Food Analysis"

Chapter 4 "EEG Measurement as a Tool for Rehabilitation Assessment and Treatment"

Chapter 5 "Fronto-Temporal Analysis of EEG Signals of Patients with Depression: Characterisation, Nonlinear Dynamics and Surrogate Analysis"

Chapter 6 "Necessity of Quantitative EEG for Daily Clinical Practice"

Chapter 7 "Periodic EEG Patterns in the Intensive Care Unit (ICU): Definition, Recognition and Clinical Significance"

Chapter 8 "Basic Electroencephalogram and Its Common Clinical Applications in Children"

I would like to express my great appreciation to the authors of this book and all those involved in its editing.

Hideki Nakano, Ph.D.
Associate Professor,
Neurorehabilitation Laboratory,
Graduate School of Health Sciences,
Kyoto Tachibana University,
Kyoto, Japan

Section 1

Basic Research of Electroencephalography

Chapter 1

EEG-Emulated Control Circuits for Brain-Machine Interface

Liljana Bozinovska and Bozinovski Adrijan

Abstract

This paper reviews efforts in a new direction of the EEG research, the direction of EEG emulated control circuits. Those devices are used in brain computer interface (BCI) research. BCI was introduced 1973 as a challenge of using EEG signals to control objects external to the human body. In 1988 an EEG-emulated switch was used in a brain machine interface (BMI) for control of a mobile robot. The same year a closed loop CNV paradigm was used in a BMI to control a buzzer. In 2005 a CNV flip-flop was introduced which opened the direction of EEG-emulated control circuits. The CNV flip-flop was used for BMI control of a robotic arm in 2009, and for control of two robotic arms in 2011. In 2015 an EEG demultiplexer was introduced. The EEG emulated demultiplexer demonstrated control of a robotic arm to avoid an obstacle. The concept of an EEG emulated modem was also introduced. This review is a contribution toward investigation in this new direction of EEG research.

Keywords: electroencephalography, EEG-emulated control circuits, brain-machine interface, robotic arm, tower of Hanoi, achievement motivation

1. Introduction

In 1929 Berger carried out research on human electroencephalogram (EEG) and introduced EEG rhythms [1]. In 1973 Vidal [2] introduced the term Brain–Computer Interface (BCI), and he set a challenge of controlling objects external to the human body by using the signals from a human electroencephalogram (EEG). He actually stated two challenges for EEG researches:

1. develop methods for EEG control of objects not being part of a human body

2. develop new methods for extracting event-related potentials (ERP) from an EEG, other than the standard averaging method.

Vidal [2] advised the use of various EEG signals, including EEG rhythms and event related potentials. Specifically, he challenged the use of the Contingent Negative Variation (CNV) event related potential.

Response to the BCI challenge was relatively slow in the years after 1973. The first report on control of an object using EEG was given by Vidal himself in 1977 [3]. He designed a responsive BCI, with active movement of eyes to elicit various visual evoked potentials (VEPs), in order to control a 2D movement of a cursor-like object on a screen. In 1988, three reports appeared related to control of objects using EEG signals. Farwell and Donchin [4] used P300 event related potential to choose a letter from a computer screen and write text on that screen. Bozinovska et al. [5–8] used contingent negative variation (CNV) event related potential to control a buzzer. That work also responded to the second Vidal's challenge, as it introduced an adaptive filter to extract a time varying CNV potential. That work introduced a new taxonomy of brain potentials [8] which classified CNV as an anticipatory brain potential. Bozinovski et al. in 1988 [9–12] used changes in EEG alpha frequency band (contingent alpha variation, CαV) to control a movement of a physical object, namely a robot; that BMI solved the long lasting challenge of telekinesis (movement of a physical object with energy emanating from a human brain). And, after the 1973 BCI challenge, it was the first intention driven (rather than response driven) BCI. Those five pioneering BCI efforts appeared before 1990. Examples of works of other authors after 1990 related to this work are [13–16].

An EEG based BCI setup consists of the following steps: (1) Produce a state in a (human or animal) brain which will be manifested by a particular EEG signal in which a control command is encoded. (2) Record the EEG signal and transmit it to a computer. (3) Analyze the EEG signal and decode the encoded command. (4) Send the decoded signal to a controlled object, such as a visual object, or a sound object, or a physical object with a mass.

EEG is a classical modality of obtaining a brain signal, but other ways of recording brain signals (e.g. magnetoencephalogram) are also being developed. This paper deals only with the EEG modality used in a BCI.

There are two ways of generating EEG-encoded commands to control an object.

1. Command encoded in an external stimulus-driven EEG response. This method uses an external stimulus (e.g., light flickering) to generate an EEG response which then is used for external object control. This way of conducting a BCI in the early 1990s was named brain-response interface (BRI) [17]. As example, a BRI directs the eye gaze toward a particular part of a computer screen which flickers in particular frequency. When a user eye looks at that part of the screen, the brain reflexively produces a corresponding visual evoked potential in the EEG, which is a function of the stimulus frequency. This method usually requires contraction of the eye muscles in order they to direct the eye gaze to a particular area on the screen, where a particular flicker is generated. An example of such method is the SSVEP (steady state visual evoked potential) method. It is a gaze-based brain-response interface method.

2. Command encoded in an intention to modulate a particular brain rhythm. This method encodes a command in EEG by an internal intention (i.e., conatively, willingly) rather than by a brain response to an external event. An example is willingly increasing the amplitude of the EEG alpha frequency band (7–13 Hz), by an idle activity (no activity), and decreasing that amplitude due to an engaging activity. For example, if the alpha band is measured from the visual area of the brain (named alpha rhythm), then closing the eyes produces a resting state and increased alpha activity. Other methods also produce relaxation and increased alpha activity. If the alpha band is measured from the motor area (named mu rhythm) then no movement and no imaginative movement produces resting state. If the alpha band is measured from the auditory area (named tau rhythm) then no-sound produces resting state.

Another term used, besides BCI (brain computer interface) and BRI (brain-response interface), is BMI (brain machine interface). It is usually used for control of an object outside of the computer screen, for example a physical object with a mass, such as a robot. With that in mind, the term BMI is used in this paper.

A BCI can be carried out invasively and non-invasively. Invasive BCI records a signal inside a brain, which requires a surgical intervention. Non-invasive BCI records EEG from the scalp, which is outside the brain. For example, the first non-invasive BMI was carried out in 1988 [9–10] for control of a (mobile) robot, and the first invasive BMI for control a robot (arm) was carried out in 1999 [18]. It is worth mentioning that those two works were the only ones in the 20th century dealing with BMI for moving object with a mass.

An essential objective of the BCI software is to find an EEG feature which can be used as a switch for controlling an object. In addition to an EEG emulated switch, recently other EEG emulated control structures are being explored, such as a flip-flop, demultiplexer, and modem. This paper will be devoted to that research.

Reviews of the BMI efforts (e.g., [19–21]) are present in the literature. Various robotic devices are being built (e.g, [22]). Many companies are involved in BMI (e.g. Emotiv [23], Kinova [24]).

In the sequel, we will first review the EEG emulated switch for control of a mobile robot. Then we will describe the EEG emulated flip-flop with applications of controlling robotic arms. Then we will describe an EEG demultiplexer and the EEG modem. Some results of current experimental research work using EEG demultiplexer are also shown.

2. EEG emulated switch

An EEG switch is a control circuit which produces a digital switch output driven by an EEG pattern. After the BCI challenge stated by Vidal [2], the first EEG switch was explicitly described in 1988 [9–12]. It is shown in **Figure 1** [11–12].

Figure 1 shows the screen of the 1988 brain-machine interface program [9–12]. The lower part of the screen shows the EEG signal recorded in a particular BMI session. Inside that session a user may generate alpha wave bursts by some relaxation technique, for example closing/opening the eyes. The robot movement takes place in real time while intentionally generating/blocking the alpha rhythm. In offline analysis mode, the program has a feature of zooming a part of the signal, defined by a line below the signal, as shown in **Figure 1**. The zoomed segment of the signal is shown in the center of the screen. The upper part of the screen shows the result of pattern recognition method in real time, which recognizes when the EEG signal contains increased amplitude of the alpha rhythm. That produces a switch pattern, actually an EEG emulated Schmitt trigger (e.g., [25]). The recognition software implements a machine learning algorithm in which the learning phase is collecting two distributions, one for amplitude and one for frequency. If both amplitudes and periods between two adjacent extreme points of an EEG signal increase, it is recognized as increased alpha rhythm, and it turns on the switch.

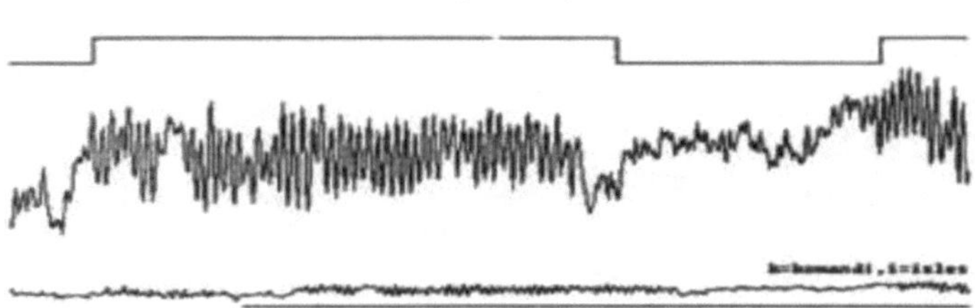

Figure 1.
EEG emulated digital switch.

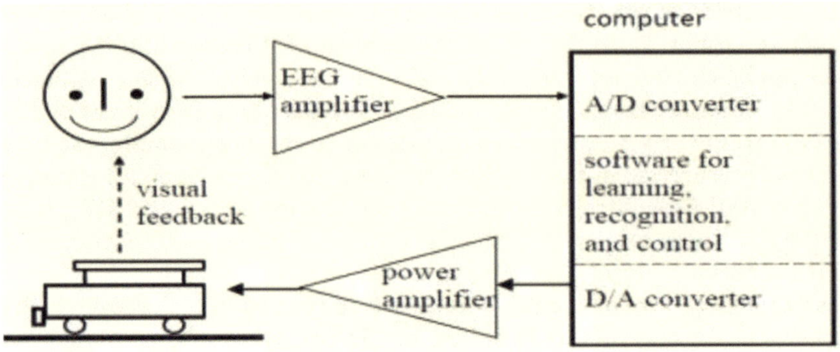

Figure 2.
The 1988 experimental setup for an EEG control of a robot.

If both amplitudes and periods decrease, it turns off the switch. A statistical machine learning method was used. Details are given in a recent review [26].

A human user, in a BCI based on an EEG emulated switch presented in **Figure 1**, requires a period of training in order to perform the task. The 1988 experiments show that the training requires about 20 minutes.

Let us note that the concept of a "mind switch" was introduced in [14]. Before that, the term "switching devices" [27] was used in relation to the independence of disabled persons. The term "brain-controlled switch" was used in [16]. We use the term EEG emulated switch as part of our work on EEG emulated control circuits.

The BMI task when the EEG emulated switch was used in 1988, was control of a mobile robot moving along a closed line drawn on the floor. Robot movement happens when the user increases the alpha rhythm by closing the eyes. The robot stops when the user opens the eyes and observes how far the robot is from the goal point, a "station" where the robot should stop. At what point to stop was decided by the user based on a visual feedback. The BMI setup is shown in **Figure 2**.

Figure 2 is a translation of the original 1988 block diagram of a BCI [9–12]. It was the first block diagram of a BCI in the literature. It shows a human user, an EEG amplifier, a computer (PC/XT), an A/D converter, a software for recognition of an alpha rhythm, a D/A converter, an energy amplifier, a robot, and an optical sensor for following the trajectory drawn on the floor. The robot used was a Ellehobby Movit Line Tracer II.

A differential biosignal amplifier was used to record the signal from the Pz site (international 10/20 system), with referential electrode on right mastoid, and ground electrode placed at the forehead. The signal was received in an IBM PC/XT (640 KB, 8 MHz) computer by an A/D converter at 300 Hz sampling rate. The software which recognized the alpha wave was written in Pascal. During the alpha wave presence, the system outputted a logic pulse at 5 volts through a D/A converter. The output signal was amplified on a transistor amplifier which drove the robot motor.

3. EEG-emulated flip-flop

A CNV flip-flop is an EEG emulation of the flip-flop digital circuit based on the contingent negative variation (CNV) event related potential (ERP). The concept of CNV flip-flop was introduced in 2005 [28].

The CNV potential [29] manifests an EEG a mental state of expectation. The CNV potential appears in an experimental procedure known as the CNV paradigm.

It is a well-known procedure (e.g., [30]) in which, in an open-loop way, a slow negative potential shift (the CNV) appears in the inter-stimulus interval of the S1-S2 stimulus pair. The negative slow potential shift is interpreted as an expectancy wave and is related to learning and memory. The classical CNV paradigm is an open-loop control system. After a stimulus S1 (visual or auditory), the brain is expecting stimulus S2 and is preparing to produce reaction R on S2. The ERP between S1 and S2 gradually develops to be a recognizable CNV. The CNV paradigm produces a ramp-like potential (the CNV) related to the pair S1-S2, but also produces a number of other evoked, cognitive, and preparatory potentials related to S1 and/or to S2.

The open-loop design for obtaining a CNV potential is given in **Figure 3**.

As **Figure 3** shows, while EEG is recorded, the user (subject) receives two sequential stimuli, by some time distance apart. The time distance between S1 and S2 (inter-stimulus interval) is fixed, 2 seconds. The time distance between S2 and next S1 (inter-trial interval) is random, 11 ± 2 seconds. Here a three - state buffer represents a trial control, where EEG enters the CNV paradigm. After several repetitions, the subject learns that after S1 follows S2 and starts to expect it. As a result, a special event related potential (ERP) appears between S1 and S2. It is a negative shift of the EEG baseline and was named Contingent Negative Variation (CNV). If a standard ERP averaging is applied, a distinctive ramp-shape potential is visible. In the classical CNV paradigm, the goal was to show the existence of a CNV potential, so the experiment ends after CNV appears, as shown in **Figure 3**.

In 1988 a feedback loop was introduced in the classical CNV paradigm [5]. In 2005 [28] it was recognized that in such a way a CNV flip-flop is emulated by an EEG. The EEG emulated CNV flip-flop is shown in **Figure 4**.

As **Figure 4** shows, a CNV flip-flop has two binary states, like an ordinary flip-flop, Q and inverse of Q (Q'). In state Q, an EEG signal is recorded as in a classical CNV paradigm, ERP is extracted after each trial, and it is tested whether the ERP is a CNV. In other words, it is tested whether the expectation *expect (S2/S1)*, is

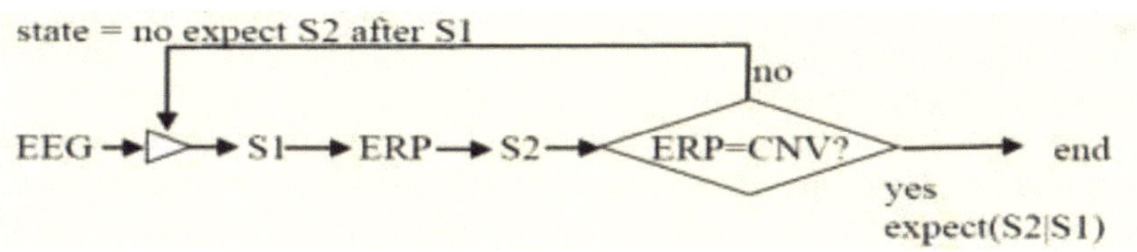

Figure 3.
The classical CNV paradigm.

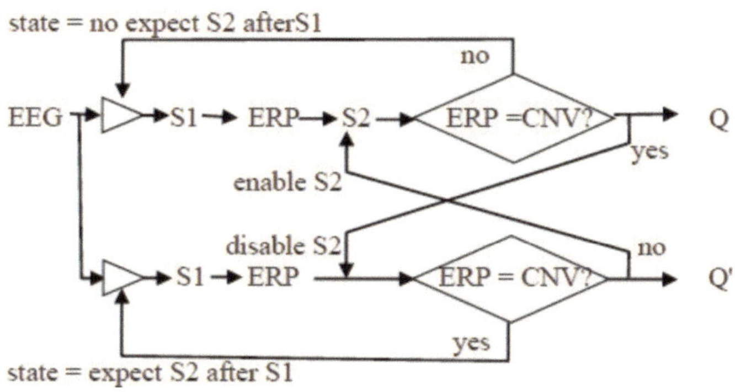

Figure 4.
An EEG emulated CNV flip-flop.

developed in the brain. However, as distinct to the classical CNV paradigm, once the CNV potential is recognized inside the recorded EEG, the flip-flop enters the sate no-Q (i.e., Q'). In this state the stimulus S2 is disabled. As a result, the CNV potential will degrade beyond recognition, which will trigger the reactivation of the S2 signal, and the flip-flop will enter the state Q again. The digital outputs Q and Q' here are used to control S2 but, can also be used to control other external devices.

Figure 5 shows the block diagram of the 1988 CNV-based BCI experiment [5–8].

As can be seen from **Figure 5**, the subject generates an EEG which contains an ERP due to the stimuli S1 and S2. The EEG undergoes initial signal processing, after which the procedure of ERP extraction follows. The final CNV pattern recognition procedure tests whether the observed ERP is a CNV. Once the presence or absence of CNV is recognized, the control signal (Enable/Disable) is sent to the controlled buzzer.

The BCI procedure starts with building CNV potential in the subject by generating S1-S2 pairs of sounds. By classical conditioning, an expectation of S2, E(S2) is being built. After repetitions, which are part of the learning process, the expectation to S2 is formed in the subject's brain, and a CNV is manifested. That event, recognition of a CNV, can be used to control an external device, such as a sound generator, a robot, or something else. In the case that expectation is not built, the CNV will gradually degrade and disappear. That point, recognition of no-CNV (no expectation) event, can also be used to control an external device, in our 1988 experiment to enable the buzzer.

A standard way of building expectation is using a reaction R(S2) to stop the duration of the S2 signal, usually by pressing a button. Pressing a button is not essential, because a subject develops expectation regardless of a motor reaction [31].

Note that the subject could willingly control the process by ceasing to build expectation, i.e., by not paying attention to the S2 stimulus. But in that case, there is no adaptive interaction between the subject and BCI, and adaptive interaction is what makes this BCI interesting.

An important feature of the CNV flip flop paradigm is that it was the first bidirectional adaptive BCI, in which both the human brain and machine are engaged to adapt to each other. This paradigm was used to study adaptive behavior in adaptive learning systems [32].

Moreover, the 1988 CNV based BCI was the first to respond to the second BCI challenge, building a method for extracting an ERP beyond the classical averaging. The need for that appeared because in the CNV flip-flop paradigm the ERP is constantly changing so the classical ensemble averaging is not applicable. An adaptive filter was needed, and the following adaptive filter was implemented.

The feature extraction module extracts the Event Related Potential (ERP). Since the paradigm requests that the obtained signal be time-variant, i.e., it forms and

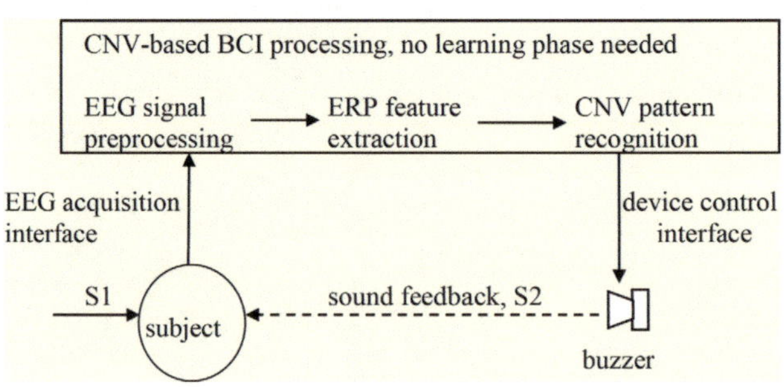

Figure 5.
The first CNV-based brain-computer interface, developed in 1988. It shows a control of a buzzer using CNV potential.

decays, a classical averaging technique is not suitable, so we used our own adaptive filter, namely

$$ERP(s, t) = pERP(s, t - 1) + qEEG(s, t) \tag{1}$$

where.

s is the sample number in a trial (s = 1,2, … , N),

t is the trial number in an experiment (t = 1,2, … ,T), and.

p and q are weighted parameters, satisfying p + q = 1.

Several parameters are collected for the current shape of the ERP, particularly important being the regression angle and the amplitudes near S1 and S2. The pattern recognition module decides whether the current ERP can be classified as a CNV. The key parameters are the slope of the regression angle and the ERP amplitude difference near S1 and S2. In the forming phase of the CNV, three consecutive confirmation trials are needed before a CNV appearance can be acknowledged.

From a practical use of a BCI, it is important that the use of a CNV flip-flop does not require separate subject training. The mental development of an expectation state is taking place in the course of the CNV experimental paradigm. In the CNV paradigm, the subject learns to expect. S/he learns that after event S1 comes event S2, and s/he adjusts her/his mental state accordingly. The mental action produces a cognitive state "after S1 expect S2."

3.1 Non-invasive BMI control of a robotic arm using a CNV flip-flop

In 2009 a BMI was built to control a robotic arm using a CNV flip-flop [33]. The task we considered was the Tower of Hanoi puzzle with two disks, the TOH(2) task.

The Towers of Hanoi, TOH(n) task, is a well known puzzle in Computer Science [34]. It has been pointed out that the solution space has a fractal structure [35]. Given a stack of n disks with different diameters, a tower is defined as a stack of disks in which the smaller disk is always above the larger one. The task is stated as follows: given three spots, A, B, and C, if the initial tower is in the spot A, move it to the spot C, using a "buffer" tower in the spot B. At each step of the task, the concept of a tower is preserved, a smaller disk always being above a larger one. It is known that to solve the TOH(n) problem, the number of required moves of disks is 2^n-1.

The BMI setup we used is shown in **Figure 6**.

The equipment used consists of a 4-channel biosignal amplifier from Biopac. The subject is connected to the biosignal amplifier with EEG electrodes placed on Cz and mastoid, while the forehead is the ground. A Windows based personal computer was used, as well as a Lynxmotion with 6-degrees-of-freedom robotic arm. We wrote the complete software in C#.

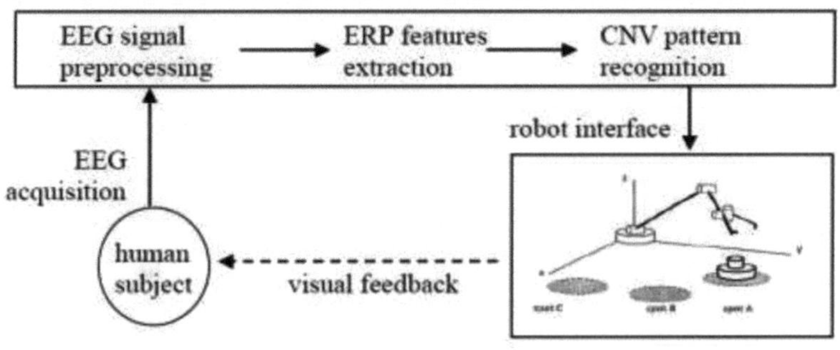

Figure 6.
BMI setup for control of a robot arm to solve TOH(2).

The preprocessing part of the software shows the obtained raw EEG and considers the EOG artifacts. The ERP extraction part extracts the ERP between S1 and S2. The CNV recognition part observes when the ERP builds a recognizable CNV, as well as when the CNV decays beyond recognition, and becomes a non-specific ERP.

The CNV flip-flop recognizes series of appearances and disappearances of the CNV potential, and triggers a behavior execution part, which moves the robotic arm toward the completion of the Towers of Hanoi task.

The robot control software receives a flip-flop signal from the CNV recognition software that a CNV is not recognizable (state Q) or is recognizable (state Q'). The flip-flop activates one of the robot behaviors stored in the memory. If there are 2 disks, i.e., the task is TOH(2), there are 2^2–1 = 3 behaviors stored. The behaviors 1 and 3 are activated by the state Q' and behavior 2 by the state Q. Each behavior is a trajectory to move a disk from current spot to the next, at a particular height. The sequence of behaviors is a solution of the TOH task. Behavior-based robotics [36] is a widely used approach in robot control.

Figure 7 is a photo of the experimental setup [37].

As **Figure 7** shows, the subject having EEG and EOG electrodes, observes the progress of TOH(2) solution, as he oscillated the state of expectation it his brain.

Figure 8 shows our graphical user interface which the experimenter observes during each trial [33].

The screen shows six rows (channels) out of which the first four are acquisition channels and the last two are mathematically computed channels. The first channel is the EEG acquisition channel, the second is the EMG acquisition from the arm pressing the button, the third is the EOG signal channel, and the fourth is the press-button recognition channel. The sixth channel is the event related potential extracted so far. If an appearance or disappearance of CNV is recognized on that channel, the signal is given to the robot to move and that is recorded on the fifth channel in **Figure 8**. In this case channel 6 shows a recognizable CNV potential, and that is signaled on channel 5. Note that CNV potential (expectancy state) is recognized before the EMG reaction signal is recognized.

TOH(2) requires 2^2–1 = 3 moves to complete the task. To see the number of BMI trials needed for solving the TOH(2) task we carried out 4 experiments and obtained results as shown in **Table 1** [33].

The data in the **Table 1** show the trial number in which the event occurred. For example, in the first experiment, the first appearance of CNV was in trial 16, the

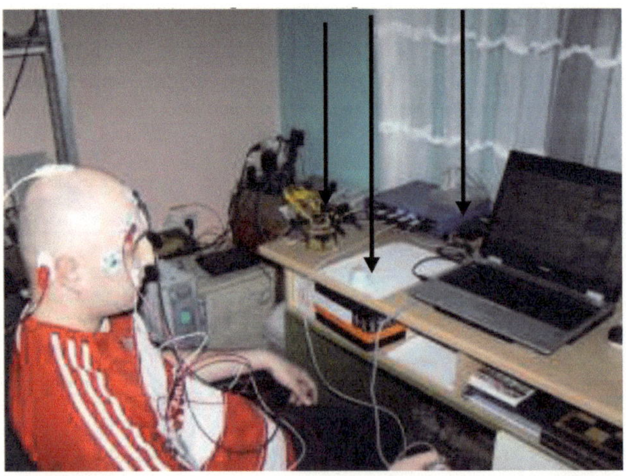

Figure 7.
Experimental setup for BMI control of a robotic arm based on a CNV flip-flop.

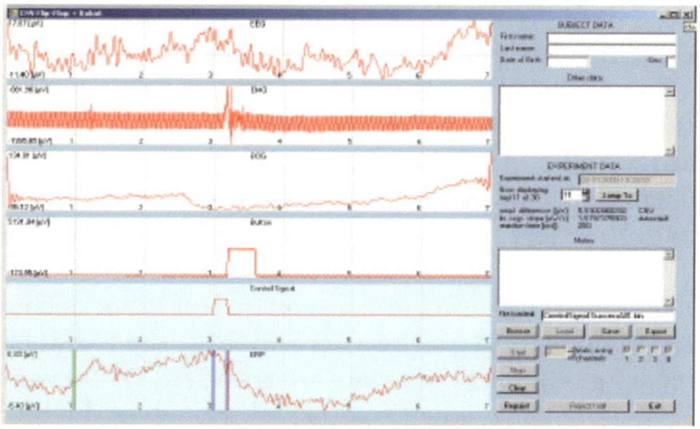

Figure 8.
An experiment trial of CNV flip-flop for robot control.

Experiment		1	2	3	4
Brain state	Robot behavior	Trial number			
CNV	behavior1: A to B	16	12	11	6
noCNV	behavior2: A to C	22	23	12	19
CNVagain	behavior3: B to C	26	29	22	22

Table 1.
Experiments of using a CNV flip-flop to control a robotic arm to solve TOH(2).

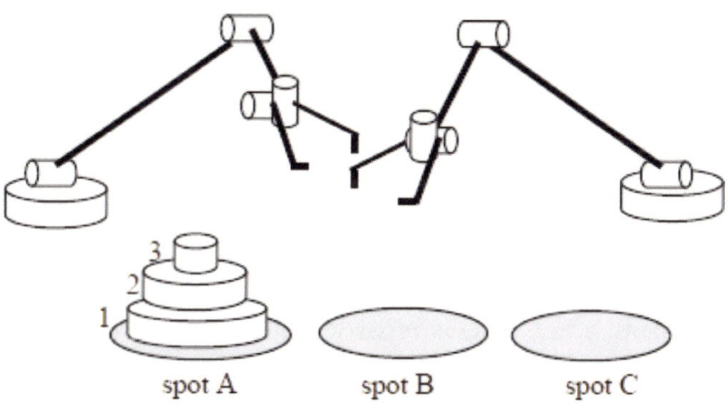

Figure 9.
The considered TOH(3) problem to be solved by two robotic arms controlled by a CNV flip-flop.

first CNV disappearance was in trial 22, and the second CNV appearance was in trial 26. As can be seen from **Table 1**, in each experiment, the two-disk Towers of Hanoi task was executed successfully within 30 trials, using this brain-machine interface.

3.2 Non-invasive BMI control of two robotic arms using a CNV flip-flop

The next BMI task considered using a CNV flip flop in a BMI setup was collaboration of two robot arms in solving the Tower of Hanoi problem with three disks, TOH(3). The task is depicted in **Figure 9** [38].

The approach is the following: Robot1 is activated by a CNV appearance event and Robot2 is activated by a CNV disappearance event. Both robots have predefined behaviors. Robots and their behaviors are triggered by a brain state recognition system, which recognizes the existence and non-existence of the brain expectancy state represented by the CNV potential.

If the height of a particular disk is denoted with a number between 1 and 3 (height 1 being the bottom), the needed sequence of robot behaviors can be defined as: A3toC1, A2toB1, C1toB2, A1to C1, B2 to A1, B1 to C2, A1 to C3. Let us note that an artificial Intelligence program was previously written for solving the general TOH(n) problem [39], where from the knowledge was used to solve this TOH(3) problem.

Once the problem is decomposed into a sequence of robot behaviors, the CNV flip-flop generates an oscillatory process that will drive the two robots with corresponding behaviors. Robot1 behaviors are activated whenever the ERP shapes into a CNV, while Robot2 behaviors are activated whenever the ERP loses its CNV shape.

In order to solve the TOH(3) task the number of moves is $2^3 – 1 = 7$. The research hypothesis for the experimental investigation is that healthy subjects will be able to carry out the oscillatory expectancy process in the brain long enough to solve the TOH(3) problem. The subject should produce the appearance of the CNV four times and the disappearance of the CNV three times. It is assumed that the TOH(3) task gives enough achievement motivation for completing the task.

The experimental setup consists of an EEG-event recognition part and a robot behavior execution part. The event recognition part recognizes the appearance/disappearance of the brain state of expectation, while the behavior execution part activates the controlled devices.

The two controlled robotic arms and the Towers of Hanoi disk set are shown in a photo in **Figure 10** [38].

Each robot is controlled by a servo controller connected to the computer by a USB-to-COM cable.

The subject is sitting and observing his/her progress toward the solution of the TOH(3) task, which gives a motivation for achievement. The EEG electrodes are placed on Cz and mastoid, while the forehead is the ground. A personal computer receives the signals and processes them. A Biopac four-channel biosignal amplifier receives the biosignal information from the subject. A USB cable connects the biosignal amplifier to the computer.

Figure 11 shows the BMI experiment screen [38].

As **Figure 11** shows, a raw EEG is recorded in channel 1, and EMG and EOG channels are recorded in channel 2 and 3. Channel 4 shows the recognized EMG and

Figure 10.
Experimental setup, two robotic arms and the TOH(3) disk set.

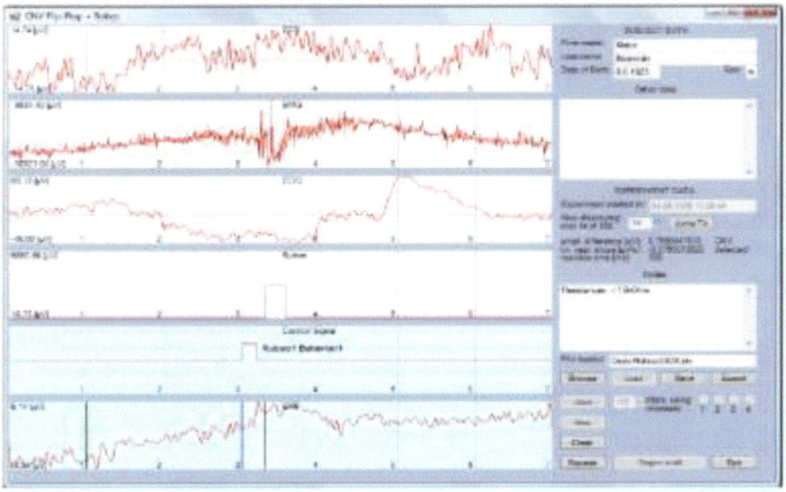

Figure 11.
The BMI experiment screen. An experimental trial. A developed CNV potential, triggers a behavior in one of the two robots.

	Experiment												
	1	**2**	**3**	**4**	**5**	**6**	**7**	**8**	**9**	**10**	**11**	**12**	**Average**
EEG event→ Robot#Behavior#	**Trial order number**												
CNV1 → Robot 1 Behavior 1 (A to C)	9	19	15	10	6	6	11	31	7	18	17	18	14
no CNV1 → Robot2Behavior 1 (A to B)	15	25	24	18	22	12	13	36	21	22	31	22	22
CNV2 → Robot1Behavior2 (C to B)	22	41	29	29	25	21	21	44	25	33	36	35	30
no CNV2 → Robot2Behavior2 (A to C)	31	45	48	40	35	32	24	50	41	40	42	42	40
CNV3 → Robot1Behavior3 (B to A)	38	60	50	52	38	36	30	55	47	46	46	45	45
no CNV3 → Robot2Behavior3 (B to C)	43	64	71	56	51	53	34	62	53	51	50	49	53
CNV4 → Robot1Behavior4 (A to C)	57	69	75	76	54	56	39	71	57	59	53	60	60

Table 2.
Results of experiments of controlling two cooperating robots by a CNV flip-flop.

channel 5 the recognized CNV signal which is sent to execute Behavior 1 in Robot1. Channel 6 shows the current ERP which is recognized as CNV.

Table 2 shows results of 12 experiments [38].

As **Table 2** shows all the experiments were successful. A human user developed her/his CNV potential in average in the 14th trial, lost it in average in the 22nd trial and so on. The task is completed in average 60 trials of a CNV flip-flop paradigm.

4. Non-invasive BMI control of a robotic arm using an EEG-emulated demultiplexer

EEG emulated demultiplexer is an emulation of a demultiplexer, a 1-to-n serial to parallel converter, a device that receives a serial input and distributes it to n outputs. An example of an EEG emulated 1-to-2 demultiplexer is shown in **Figure 12** [40]. It is driven by alpha rhythm.

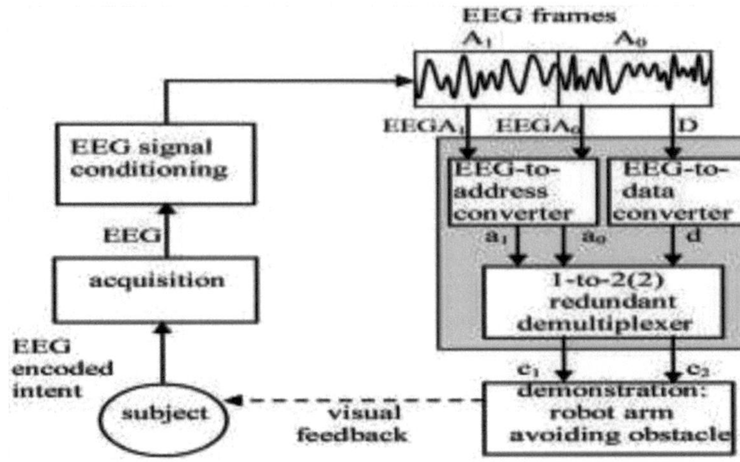

Figure 12.
EEG emulated demultiplexer.

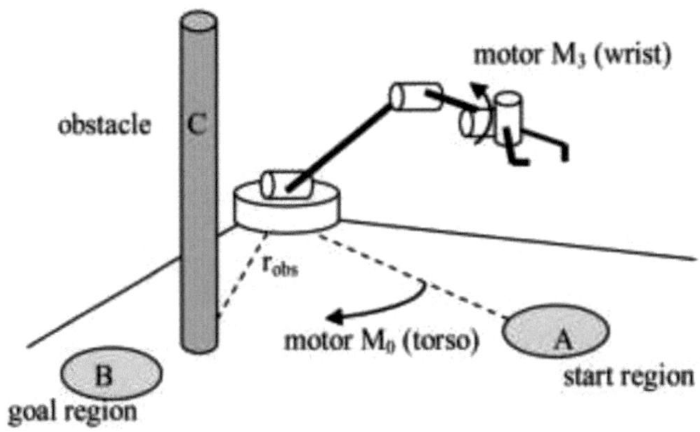

Figure 13.
The BMI task of controlling two motors by a single EEG channel using EEG demultiplexer.

As **Figure 12** shows, a serial EEG signal is divided into two segments (frames). Those two EEG frames, A_0 an A_1, with encoded intent for commands to external devices, are sent to the EEG demultiplexer. Both frames can be defined as binary channels, but A_0 can also be defined as multi-valued channel. The binary channels are used as binary addresses for the address decoder of the demultiplexer, while the multivalued channel is viewed as a data input of the demultiplexer. The EEG demultiplexer contains an address converter and a data converter. The signals enter the redundant 1-to-2 demultiplexer which sends the data channel to one of the two output channels c_1 and c_2, defined by the address decoder. Those output channels control two devices, for example two motors of a robotic arm. The demultiplexer used is redundant, because for addressing 2 output channels it uses 2 address lines, instead of just one. That has been done to increase the accuracy of EEG addressing (e.g., [41]).

The BMI task considered is shown in **Figure 13** [40].

As shown in **Figure 13**, a robotic arm should move from start region A to goal region B. The horizontal projection of the arm is such that if moved toward goal area B, it would hit an obstacle C along the way. In order to avoid the obstacle, the wrist of the arm should be moved up, so that horizontal projection of the arm is shortened before it reaches the obstacle C. The BMI task for the subject is: from a single EEG channel, generate an EEG pattern that will move the robot arm from A to B, avoiding C. So, the task is to control two motors from a single channel EEG.

Figure 14 shows the experimental setup [40].

As **Figure 14** shows, a subject is sitting in front of a robotic arm and sends EEG commands such that the task of moving the arm while avoiding and obstacle is achieved.

The experimental trial of an EEG demultiplexer controlling a robot is shown in **Figure 15** [40].

As **Figure 15** shows, a raw EEG is received by the BMI system and is shown in Channel 1. Channels 2 and 3 are not recorded. Channel 5 is the filtered EEG to obtain the alpha rhythm. Channel 4 is the filtered alpha rhythm to obtain a signal which represents the alpha rhythm envelope. That signal is tested against a threshold value, shown in the same channel. Channel 6 contains two frames, each showing

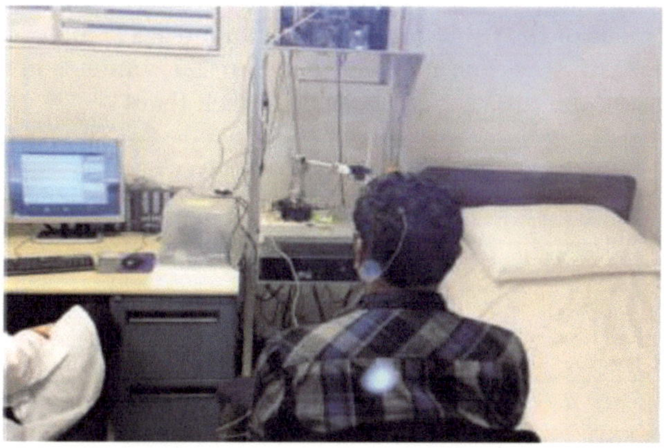

Figure 14.
Experimental setup for a BMI based on a EEG demultiplexer.

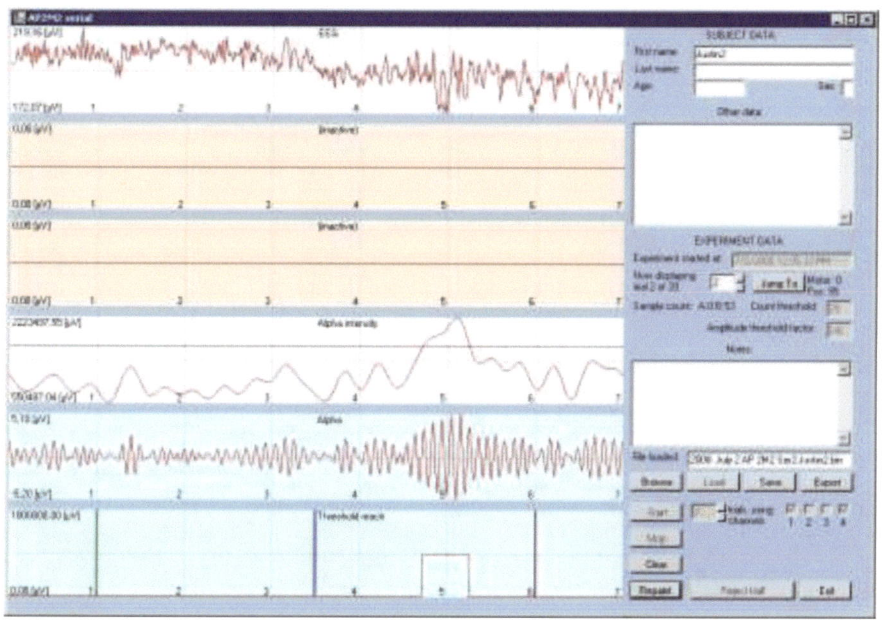

Figure 15.
The screen of an experimental trial of an EEG demultiplexer.

a pulse for how long the duration in which the envelope is above the threshold. Also, in the **Figure 15** can be seen that the binary value of frames is $A_1A_0 = 01$.

The channel 6 is the EEG demultiplexer channel. First the binary values of the frames are computed, in this case $A_1A_0 = 01$. That is a command to send the data to the chosen motor. The data are computed from the duration of the signal in frame A_0, and a signal to move is sent to the motor. The demultiplexer commands are defined as $A_1A_0 = 00$ do nothing, $A_1A_0 = 1X$, change motor, and $A_1A_0 = 01$ move motor. Thus, control of two motors using a single channel EEG is achieved.

Table 3 [40]. shows an experiment of a BMI using EEG demultiplexer in solving the problem of moving a robotic arm from A to B avoiding an obstacle at point C along the way.

As can be seen from **Table 3**, the threshold value of the alpha band envelope is set to 25. At trial 1, the frame A_1 has a value of $20 < 25$, and frame A_0 has a value $0 < 25$. So the binary values of the input lines to the demultiplexer are $a_1a_0 = 00$. The output line of the demultiplexer is c_1, which activates motor M_0 which is for horizontal movement of the robot arm. The command $a_1a_0 = 00$ means "do nothing" and the robot arm stays and its initial position 127, which is in the start region A. In the second trial the subject generates EEG such that $C_1 = 23 < 25$, and $C_0 = 36 > 25$, so the input demultiplexer lines are $a_1a_0 = 01$. The currently addressed

Trial	EEG Demultiplexer					Robot		
	Threshold $\theta_c = 25$				Output line	Motors		
	$a_k = \text{sgn}(C_k - \theta_c)$					Motor	Command	Position
	C_1	a_1	C_0	a_0				
1	20	0	0	0	c_1	M_0	NoOP	127
2	23	0	36	1	c_1	M_0	Move	112
3	1	0	88	1	c_1	M_0	Move	89
4	5	0	45	1	c_1	M_0	Move	76
5	0	0	62	1	c_1	M_0	Move	60
6	0	0	28	1	c_1	M_0	Move	53
7	23	0	31	1	c_1	M_0	Move	39
8	77	1	4	0	c_2	M_3	Switch	127
9	11	0	43	1	c_2	M_3	Move	138
10	0	0	47	1	c_2	M_3	Move	150
11	0	0	35	1	c_2	M_3	Move	159
12	0	0	44	1	c_2	M_3	Move	170
13	2	0	55	1	c_2	M_3	Move	184
14	0	0	75	1	c_2	M_3	Move	203
15	0	0	65	1	c_2	M_3	Move	220
16	0	0	65	1	c_2	M_3	Move	237
17	25	1	17	0	c_1	M_0	Switch	39
18	0	0	19	0	c_1	M_0	NoOP	39
19	0	0	54	1	c_1	M_0	Move	25

Table 3.
Experiment of a BMI using EEG demultiplexer to control a robotic arm to move from point a to point B avoiding an obstacle at point C along the way.

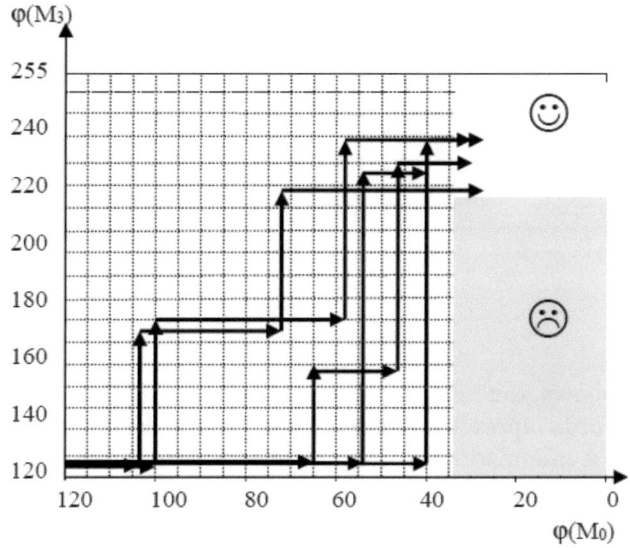

Figure 16.
Achievement motivation space for experiments in BMI for controlling a robot arm using an EEG demultiplexer.

motor M_0 moves from position 127 to position 112. The subject drives the robot arm horizontally, up to position 39. The obstacle is at position 35, so the subject changes the movement to the motor M_3 which will move the arm wrist vertically. It should be noted that the subject does not know the internal coordinates of the motors, and s/he only sees the movement in space, and s/he estimates how far the robotic arm is from the visible obstacle. In trial 8 the subject changes the alpha rhythm pattern co that $A_1 = 77 > 25$ and $A_0 = 4 < 25$, i.e., $a_1a_0 = 10$ which changes the demultiplexer output and chooses the motor M_3 which is still in its initial position 127. In trial 9 s/he moves that motor to position 136. With careful BMI control, the subject succeeds to achieve the goal area in robot coordinates $M_0M_3 = (25, 217)$, avoiding the obstacle at $M_0M_3 = (< 34, <217)$. Any position of $M_3 < 217$ would hit the obstacle at $M_0 < 34$.

Figure 16 [42] shows some results of the experiments of a BMI using EEG demultiplexer in controlling a robotic arm, as described above.

The experimental investigation carried out 53 BMI experiments. Successful were 42 of them. **Figure 16** shows example of 5 experiments. Here the goal region is marked with symbol ☺ and the avoidance region (obstacle) with symbol ☹. The participants build behavioral trajectory through the achievement motivation space in order to reach the goal region while avoiding the obstacle. The coordinates are the internal robot coordinates, unknown to the participants. The participants use the view of the robotic arm to navigate the arm using their EEG.

As can be seen from **Figure 16**, the starting region of robot movement in each experiment is around the coordinate $M_0M_3 = (127,127)$. Using BMI and controlling generated alpha rhythm in an EEG sentence, various trajectories are achieved toward the goal region M_0M_3 (>34, >217), avoiding the obstacle region $M_0M_3 = (< 34, < 217)$.

5. EEG emulated modem

An EEG emulated modem [40] is a process in which a sentence (message, command) is encoded in an EEG and is decoded at some receiving site, for example in a computer. **Figure 17** shows the concept.

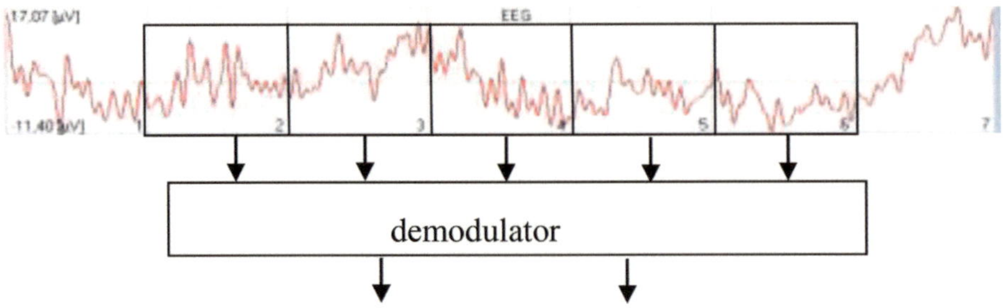

Figure 17.
EEG emulated modem.

As **Figure 17** shows, the EEG signal is viewed as an EEG encoded sentence, which contains words represented by EEG frames. The sentence is encoded as an EEG modulation. A modulation process usually contains a carrier signal which is a good harmonic signal, modulated by a message. The EEG carrier signal is a stochastic (or chaotic) signal [43], and it has some statistical properties, such as mean value and standard variation, among others. And it can be decoded given some information about the encoding process. For example, if it is known that the message is encoded in the alpha band, then first the alpha band can be filtered out, and the envelope can be obtained containing the message, as it was done in [40].

Here the concept of modulation is wider than the classical harmonic signal modulation. It can be any way of encoding a sentence in an EEG.

The EEG modem is an approach toward application of BCI with a low number of channels, when several devices should be controlled with a minimum number of EEG channels.

6. Conclusion

EEG emulation of control circuits is a new direction in EEG research. It was introduced in 2005 with the concept of a CNV flip-flop. However, after the Vidal's BCI challenge in 1973, the first explicit description of an EEG emulated electronic circuit was given in 1988. It was an EEG emulated switch, actually a Schmitt trigger, based on the EEG alpha rhythm. Recently in 2015 the EEG emulated demultiplexer and EEG emulated modem were described. This paper is a first review of this new direction in EEG research.

Acknowledgements

This research is supported in part by the current grant USDA/NIFA SCX-311-23-17 to the first author. Part of this research was supported by Macedonian-Croatian bilateral research grant to both the authors in the period 2007-2010. The authors would like to express gratitude to Drs. Stanko Tonković, Velimir Išgum, and Ratko Magjarević for their help.

Author details

Liljana Bozinovska[1]* and Bozinovski Adrijan[2]

1 South Carolina State University, Orangeburg, USA

2 University American College Skopje, Macedonia

*Address all correspondence to: lbozinov@scsu.edu

References

[1] Berger, H. Ueber das Elektrenkephalogramm des Menschen. *Archiv fuer Psychiatrie Nervenkrank,* 1929; 87: 527–570.

[2] Vidal, J. Toward direct brain-computer communication. *Annual Review of Biophysics and Bioengineering*. 1973; 157–180.

[3] Vidal, J. Real-time detection of brain events in EEG. *Proceedings of the IEEE.* 1977; 65: 633–641.

[4] Farwell, L, Donchin, E. Talking off the top of your head: a mental prosthesis utilizing event-related brain potentials. *Electroencephalography and Clinical Neurophysiology.* 1988; 70: 510–523.

[5] Bozinovska, L, Sestakov, M, Stojanovski, G, Bozinovski, S. Intensity variation of the CNV potential during the biofeedback training guided by a personal computer [abstract] (In Serbian). *Neurologija*supppl 2. 1988; 37: p. 76.

[6] Bozinovska, L, Bozinovski, S, Stojanov, G, Sestakov, M. Introduction of biofeedback in the CNV paradigm. (In Serbian), In: Proceedings Conference ETAN, Novi Sad, 1989; XII: 93–98.

[7] Bozinovska, L, Stojanov, G, Sestakov, M, Bozinovski, S. CNV pattern recognition - A step toward cognitive wave observation, In: L. Torres, E. Masgray, M. Lagunas (editors). Signal Processing: Theories and Applications. Proceedings of Fifth European Signal Processing Conference (EUSIPCO 90); Barcelona. Elsevier Science Publishers, 1990; 1659–1662.

[8] Bozinovska, L, Bozinovski, S, Stojanov, G. Electroexpectogram: Experimental design and algorithms, In: Proceedings IEEE International. Biomedical. Engineering Days. Istanbul, 1992; 58–60.

[9] Bozinovski, S, Sestakov, M, Bozinovska, L. Control of a mobile robot using alpha rhythm from a human brain, (In Croatian), Proceedings Symposium JUREMA, Zagreb, 1988; 247–249.

[10] Bozinovski, S, Sestakov, M, Bozinovska, L. Using EEG alpha rhythm to control a mobile robot. In: G. Harris, C. Walker (editors), Proceedings 10th Annual Conference of the IEEE Engineering in Medicine and Biology Society, track 17, Biorobotics, New Orleans, 1988; 10: 1515–1516.

[11] Bozinovski, S, Sestakov, M, Stojanov, G, Bozinovska, L. Bioelectric control of mobile robots, (In Macedonian) Proceedings 6th Yugoslav symposium for applied robotics and flexible automation, Novi Sad, 1989; 237–242.

[12] Bozinovski, S. Mobile robot trajectory control: From fixed rails to direct bioelectric control, In: O. Kaynak (Ed.) Proc. IEEE International Workshop on Intelligent Motion Control, Istanbul, 1990; 2:463–467.

[13] Keirn, Z, Aunon, J. Man-machine communication through brain-wave processing. *IEEE Engineering in Medicine and Biology Magazine*, 1990; 55–57.

[14] Craig, A, Kirkup, L, McIsaak, P, Searle, A, The mind as a reliable switch: Challenges of rapidly controlling devices without prior learning. In S. Howard, J. Hammond, G. Lindgaard, (Eds.) *Human Computer Interaction*. London, Chapman and Hall. 1997; 4–10.

[15] Searle, A. Electrode performance and signal processing strategies for discrimination of EEG alpha waves: Implementations for environmental control by unconstrained subjects without training. PhD. Thesis,

Department of Applied Physics and Technology, Sydney, 2000.

[16] Mason, S, Birch, G. A brain-controlled switch for asynchronous control applications, *IEEE Trans. Biomedical Engineering* 2000; 47 (10): 1297–1307.

[17] Sutter, E. The brain response interface: communication through visually induced electrical brain responses. *Journal of Microcomputer Applications*. 1992; 15: 31–45.

[18] Chapin, J, Moxon, K, Markowitz, R, Nicolelis, M. Real-time control of a robot arm using simultaneously recorded neurons in the motor cortex, *Nature Neuroscience*. 1999; 2: 664–670.

[19] Lebedev M. Augmentation of sensorimotor functions with neural prostheses. *Opera Medica and Physiologica*. 2016; *2* (3):211–227.

[20] Lebedev, M, Nicolelis, M. Brain-machine interfaces: from basic science to neuroprostheses and neurorehabilitation. *Physiological Review*.2017; 97: 737–867.

[21] Liu, X, van der Spiegel, J. *Brain Machine Interface*. Springer Verlag. 2017.

[22] Li, W, Jaramilo, C, Li, Y. A brain computer interface based humanoid robot control system. In Proc. IASTED Int. Conf. Robot., Pittsburgh, 2011; 390–396.

[23] Emotiv. The introductory guide to BCI (Brain-Computer Interface). [Internet]. 2020; https://www.emotiv.com/bci-guide/

[24] Ledoux, M. How existing assistive HMIs could change our near future: Part II [Internet] 2020; https://www.kinovarobotics.com/en/knowledge-hub/how-existing-assistive-hmis-could-change-our-near-future-part-ii

[25] Sherz, P, Monk, S. *Practical Electronics for Inventors*. McGraw-Hill, 2013.

[26] Bozinovski, S, Bozinovska, L. Brain-computer interface in Europe: The thirtieth anniversary. *Automatika*. 2019; 60(1): 36–49.

[27] Barnes, M. Switching devices and independence of disabled people. *British Medical Journal* 1994; 309: 1181–1182

[28] Bozinovski A. CNV flip-flop as a brain-computer interface paradigm. Proceedings 7th Symposium on Medical Informatics. Rijeka, Croatian Society of Medical Informatics, 2005;7:149–154.

[29] Walter, G, Cooper, R, Aldridge, V, McCallum, W. Contingent negative variation: An electric sign of sensory-motor association and expectancy in the human brain. *Nature*. 1964; 203: 380–384.

[30] Tecce, J, Cattanach, L. Contingent Negative Variation (CNV). In E. Niedermeyer, F. Lopes da Silva (eds.) *Electroencephalography: Basic Principles, Clinical Applications, and Related Fields*. Urban and Schwarzenberg, 1982; 543–562.

[31] Bozinovski A, Tonkovic S, Isgum V, Bozinovska L Electrophysiology of the expectancy process: Processing the CNV potential. Proceedings of the 5th International Conference on Informatics and Information Technology. Bitola. 2007; 129–137.

[32] Bozinovski S., Bozinovska L. Anticipatory brain potentials: An electrophysiological insight into the anticipatory behavior of adaptive learning systems. In M. Butz, O. Sigaud, S, Swarup (Eds.) Proc. Workshop on Anticipatory Behavior of Adaptive Learning Systems. Los Angeles, 2004; 1–10.

[33] Bozinovski, A, Bozinovska, L. Anticipatory brain potentials in a brain-robot interface paradigm. Proc. 4th International IEEE EMBS Conference on Neural Engineering. Antalya. 2009; 451–454.

[34] Meissner, L. *The Science of Computing*. Waldsworth Publishing Company, 1974.

[35] Bozinovski, S. *The Artificial Intelligence* (In Macedonian) Gocmar Press, 1994.

[36] Arkin, R. *Behavior-Based Robotics*, The MIT Press, 1998.

[37] Bozinovski, A. Brain-machine interface - A 21st century dynamic technology: Anticipatory brain potentials and robot control. Proc of the Global Forum: Shaping the Future, Washington, DC, November 8–9, 2010

[38] Bozinovski, A, Tonkovic, S, Isgum, V, Bozinovska, L. Robot control using anticipatory brain potentials, *Automatika*. 2011; 52(1):20–30.

[39] Bozinovski, A. Control software for the Towers of Hanoi problem solving, using a manipulative robot – A case study, Proc. 2nd Balkan Confer4ence on Informatics. Ohrid. 2005; 366–378.

[40] Bozinovski, S, Bozinovski, A. Mental states, EEG manifestations, and mentally emulated digital circuits for brain-robot Interaction. *IEEE Transactions on Autonomous Mental Development*. 2015; 7(1): 39–51.

[41] Kuekes, P, Robinett, W, Seroussi, G, Williams, S. Defect-tolerant interconnect to nanoelectronic circuits: Internally redundant demultiplexers based on error-correcting codes. *Nanotecnology*. 2005;(16):1419–1432.

[42] Bozinovska, L, Bozinovski S. An achievement motivation space for brain-robot interface. Proc. IEEE SoutheastCon, Huntsville, AL. 2018

[43] Dmitriev, A, Kuzmin, L, Laktushkin, A. Amplitude modulation and demodulation of chaotic carriers. In Proc. 12th IEEE Workshop on Nonlinear Dynamics in Electronic Systems. (NDES'2004), Evora. 2004;138–141.

Chapter 2

EEG Analysis during Music Perception

González Almudena, Santapau Manuel
and González Julián Jesús

Abstract

This review presents the most interesting results of electroencephalographic studies on musical perception performed with different analysis techniques. In first place, concepts on intra-musical characteristics such as tonality, rhythm, dissonance or musical syntax, which have been object of further investigation, are introduced. Most of the studies found use listening musical extracts, sequences of notes or chords as an experimental situation, with the participants in a resting situation. There are few works with participants performing or imagining musical performance. The reviewed works have been divided into two groups: a) those that analyze the EEGs recorded in different cortical areas separately using frequency domain techniques: spectral power, phase or time domain EEG procedures such as potentials event related (ERP); b) those that investigate the interdependence between different EEG channels to evaluate the functional connectivity between different cortical areas through different statistical or synchronization indices. Most of the aspects studied in music-brain interaction are those related to musical emotions, syntax of different musical styles, musical expectation, differences between pleasant and unpleasant music and effects of musical familiarity and musical experience. Most of the works try to know the topographic maps of the brain centers, pathways and functions involved in these aspects.

Keywords: EEG, network-graph, phase synchronization, functional connectivity, music

1. Introduction

The literature on the use of EEG analysis in musical perception processes engage very different aspects related to the effects and processes of music in the brain, such as musical emotion, cognition, musical syntax, etc. In this review at the introduction, we summarize different aspects of the characteristics of musical styles to afterwards develop the EEG applications in different music-brain interactions.

Music is an art present in our daily life through numerous styles and forms. We know that musical perception involves different human factors, on the one hand it produces emotion and sensations, and it is also related to syntactic processing [1–4]. Musical perception is also modulated by factors associated with the personal characteristics of the listener: age, cultural level, socioeconomic and cultural context in which they live, musical experience and learning, familiarity with the type of music hearing, psychological state and preferences [5–7]. On the other hand, the dichotomy between cognition and musical sensation has been widely studied; the literature finally proposes a continuum between the both concepts [8]. The reception of musical content not only consists of sensations (e.g. happiness or melancholy) it also implies aspects such as recognition of structures and/or predictability of musical discourse. For all this, it is a mixture of concepts that leads to cognitive and sensitive perception.

When talking about musical perception, the importance of tonality (previously modality), should be highlighted, which is known to have been present in Western academic and popular music since medieval times. Harmony is the structure from which the tonal system departs. This structure consists of intervallic relations that can be consonant, perceived by humans as pleasant, or dissonant, perceived as unpleasant or at least as a moment of tension. In musical syntax consonant chords tend to be associated with relaxation, and dissonant with tension. Music theory, and harmony in particular, is a very extensive and complex field of knowledge that has been transformed by musical styles and the evolution of history, hence our simplified summary of intervallic relations. Different fields of knowledge such as the electroencephalography literature have analyzed many of these concepts.

The cognitive component in music recognition has been of great importance. So, musical cognition is related to the use of long-term memory, since it refers to the perceptual and to what the brain has learned in terms of the tonal hierarchical structure overwhelmingly present in our culture [6]. The sensory is related to short-term memory, to how we receive the sound result of the tonal distributions in Western music [9]. Another very important issue that has been extensively dealt is the predictability of the tonal system. Structurally, tonal music evolves temporarily within specific tonal/spectral ranges and with relatively low uncertainty (entropy) limits that make it reasonably predictable. Recent theories along these lines, such as that of McDermot, analyze familiarization with the tonal system of society against theories of auditory neurobiology underlying the attraction factor to the tonal structure [10]. This aspect of neurobiology has been discussed by Bowling et al., who analyze this issue and conclude that without the exposure of the general population to the system or tonal structure, it is undeniable that there is underlying biological evidence that demonstrates that both concepts are closely linked [11].

In addition to tonality, concerning Western academic music also contains defined rhythmic forms. This means that when listening, we can recognize patterns that develop temporarily (or within a sequence of time) and that lead to a continuous generation of expectations and predictions [12–14] together with a certain capacity for anticipation [15]. These three concepts: expectation, prediction, and anticipation are closely linked to our perception of music and occur without us being aware of them. Two main sources of musical expectations have been described: the explicit knowledge of how a piece of music with which one is familiar will develop and the implicit understanding that comprises the knowledge of the rules of music while listening [16]. The implicit expectations arise because each musical style, genre, and culture contains specific rules, patterns, sound characteristics, and time. Exposure to music training or social and cultural influences affect and determine an individual's emotional response [17].

Musical preferences are also believed to be due to the completion of the expectation of a pattern [18], that is, whether the expectation is met positively or negatively. When listening to music, –for regular listener- the listener expects certain patterns of notes or phrases, and this entails a prediction of the musical event [12], which can be frustrated in some musical styles like contemporary/new music. Zatorre's review, identifies some of the auditory cortical circuits responsible for coding and storing tonal patterns, and discusses evidence that shows the importance of cortical loops between auditory and frontal cortices for maintaining musical information in working memory and for the recognition of structural regularities in musical patterns that then lead to expectations [19]. In the tonal system, the emotional effects of the alteration of predictions include surprise. Irregular or unexpected chord types evoke skin conductivity responses, and the range of such emotional responses is related to the degree of surprise at the unexpected [20]. Therefore, the three cognitive aspects -expectation, prediction and anticipation- that occur when listening to tonal music are also related in generating musical emotions.

The different aspects of musical perception mentioned above have been investigated since the last century by means of different types of analysis and techniques based on electroencephalographic signals. In fact, it has been considered that the EEG frequency oscillations are crucial to link different elements and merge them into a coherent perception something relevant for the processing of music considered as a multifunctional stimulus [21]. There is a review of the neural bases of musical perception by Koelsch in which different signal and neuroimagen techniques are considered including some based on EEG signals [22].

2. Electroencephalographic (EEG) signals analysis during music perception

The first quantitative analyzes of brain activity using EEG signals date back to the 70s of the 20th century after the appearance of the fast Fourier transform that allowed the representation the EEG signal spectral power in different frequency bands. Specifically, electrical brain activity has been used in brain music research using univariate procedures, that is, analyzing the individual activity of EEG/MEG (magneto-EEG) /ERP (event related potentials) signals extracted from certain cortical brain areas/channels. Multivariate procedures have also been used where interdependence, correlation or synchronization between two or more channels are evaluated (see **Figure 1**).

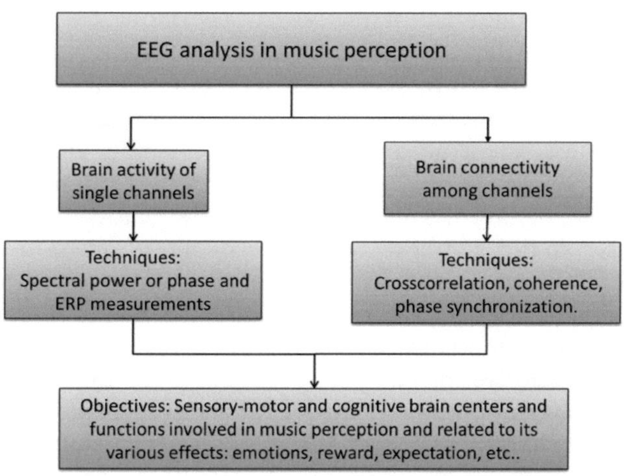

Figure 1.
Summary block diagram of the sections included in the EEG review.

2.1 Single channel analysis

2.1.1 General effects of listening music

EEG spectral power measurements from different cortical areas appear to indicate that musical processing may entail local and/or distant neural networks whose communication may affect different EEG frequency bands [23], such as changes in alpha power in the parieto/occipital and fronto/temporal regions [24], in beta power in the right parietal/temporal cortex [25], or in gamma power in the right parietal region [26]. Also, depending on the type of tonal music heard, different alterations in EEG spectral power occur in different bands [27, 28], located bilaterally in certain cortical areas [29].

2.1.2 EEG and musical emotions

Others studies have examined changes in spectral power in different EEG frequency bands at different brain regions as indicators of musical emotion processing. Concerning music emotions, in a review by Koelsh about music processing, it is reported that sounds are structured in time, space, and intensity, and that the perception of musical structures has emotional effects, which emerge from the music itself [20]. This occurs due to the processing of the intra-musical structure and the concept of musical tension. This author, gives an extensive explanation of the underlying structural factors that give rise to this concept and that nowadays we can observe/identify in many musical styles, including some as current as pop or rock. On the other hand, musical emotions are closely related to the concept of musical familiarization, it is worth highlighting the familiarization with the tonal system in which practically all the music that we perceive are immersed and that we have mentioned above. In this line with the image analysis technique (FMRI) familiar music appears to activate the limbic and allocortex systems, and areas associated with the reward mechanism: areas of the cingulate and frontal lobe, which in turn are not activated with unfamiliar music [5]. Also, brain areas in the right striatum and the orbitofrontal cortex have been related to specific emotions such as joy [30] and music that produces happiness increases activity in the striatum, cingulate, and posterior Heschl's gyri, while sad music activates the anterior hippocampus and the amygdala [31]. As well areas reported to be activated in the reward system in musical emotion are the ventral striatum, the insula, and the orbitofrontal cortex [32]. It is a fact that most people derive pleasure from music. Hearing especially expressive musical stimuli is reported to evoke emotion and neuronal activations relating to the reward system. It also produces an affective impact on the listener's brain, which can be altered by the subjects' musical training [33]. In other words, familiarity with music affects perception of it. Zatorre considers the evidence on how the mesolimbic striated system participates in reward, motivation, and pleasure in other domains [19]. In this line, several authors consider that the areas associated with emotions and reward is also involved in the emotional response to music [34, 35]. Limbic and paralympic areas respond to the dynamic expression of the musical interpretation of humans [33], specifically accord to tonal music, the cingulate subcallosal gyrus, the anterior prefrontal cingulate, the retrosplenial cortex, the hippocampus, and the anterior insula [36]. Therefore, through the analysis of neural networks, the literature suggests that the subcortical dopaminergic regions work in conjunction with important cortical regions to give rise to esthetic pleasure. These regions are related to the reward system, a set of structures that, through stimuli, in this case the auditory, provide humans with pleasure or can modify behaviors through positive reinforcement.

Several studies have analyzed increases or decreases in power in different bands when listening to pleasant or unpleasant music [26, 37]. Furthermore, it has been reported that pleasant as opposed to unpleasant music appears to increase the strength of the frontal midline theta band [38]. On the other hand, the hypothesis of asymmetry has been postulated, which proposes that positive and negative emotions are processed mainly in the left and right frontal brain regions, respectively [39]. Various measures of asymmetry, not necessarily in the frontal cortex, have been used to develop quantitative tools to assess emotions caused by visual [40] and musical [41] stimuli. It has also been reported lateralization of EEG activity in the alpha band due to opposite valences and different cortical topographies of lateralized alpha activation have been found for different musical patterns [28, 42]. In addition, theories about the cortical topography of musical emotion where the hemispheres would have specialized functions have been studied [43].

Another aspect that can modulate musical emotions is the musical familiarization which is normally closely linked to musical genres. Thus, bilateral frontal-temporal alteration in EEG spectral power has been reported while the subjects listened to musical extracts of different genres -structure-tonal or environmental origin- [29]. Moreover, the preferences of the subjects towards different musical extracts have also been studied, using characteristics extracted from the frequency-time analysis of the EEG signals. Once the listeners' familiarity with the excerpts was considered, the classification accuracy increased for familiar music [44]. Another point of interest is the musical concept of dissonance that has been widely discussed in the literature. Dissonance can be part of harmonic language creating musical tension and is in fact common in musical languages until the 20th century. But dissonance has also been studied as an isolated concept and related as an uncomfortable or unpleasant sound in some relation to noise. Thus, several centuries of the established use of dissonance to create unexpected and disconcerting moments is altered, the general EEG activity recorded in the left hippocampus has been reported to discriminate changes from consonance to dissonance [45], said of another as from the consonance of musical intervals to the dissonance of, for example, a major or minor second [46]. Also using PET images, it has been reported that the gradual variation from consonance to dissonance is accompanied by a gradual decrease in neuronal activity in some cortical areas (orbital and ventromedial prefrontal and subcallosal cingulate) but increases in other subcortical areas such as the parahippocampal and precuneus gyri [34]. Evidently, if we grow up and develop in a cultural environment of tonal music, our neuronal brain centers relating to emotion and musical cognition will adapt. For this reason, the sounds that make up a piece of music can be considered esthetically pleasing or not, depending on acoustic properties such as the use of harmony within that system [47] which can give pleasure to the listener or not.

2.1.3 EEG and music styles and experience

The style of music heard and its intramusical characteristics show different alterations in the spectral power of the EEG in different bands [27, 28]. On the other hand, there are research papers that reveal the impact of musical experience on musical brain processing. Thus, using MEG, it has been found signal oscillations phase blocking in gamma improved in expert musicians versus non-musicians during audition of dissonant and minor chords [48]. Moreover, in an EEG study carried out with expert musicians, in this case saxophone players playing in ensemble (quartet), found alterations in power potency in brain areas BA44/45 involved in semantic functions [49]. It therefore seems evident that musical experience is an important condition that intervenes in musical processing by the brain.

2.1.4 ERP analysis in brain musical processing

There is abundant literature in relation to musical syntax studies by using the EEG event related potentials (ERP) at different cortical areas. Thus, measuring negative/positive ERP peaks latencies syntactic language and harmony incongruities has been investigated [50] or whether language and music processing share processing resources: both appear to activate non-identical syntactic connections [51] and also has been reported how the two forms of music expectations –explicit and implicit that we explained above- manifest themselves with different neuronal correlations [52]. In ERP experiments where repetitive auditive stimulation was produced, early right anterior negativity (ERAN) has been found [53]. In unpredictability experiments where the position of the irregular chords is unknown, that is, when the musical expectation is broken in a sequence of sounds, the negativity usually has a longer latency and an anterior-temporal distribution (RATN) [54]. Additionally, analysis of incoming harmonic sequences elicited an early effect, taken as the magnetic equivalent of the ERAN (termed mERAN) localized in Broca's area and its right-hemisphere [1]. It has also been shown with this kind of analysis that when listening to melodies with irregular tones, the early right anterior negativity has a shorter maximum latency than that caused by irregular chord functions [55]. Therefore, a difference in musical perception in relation to musical expectation has been demonstrated through different paradigms of syntactic irregularities in chords or melodies. However, in other ERP studies on syntactic processing of music and language report shared neural resources, or what is the same, interactions between music-syntactic and language-syntactic processing [56, 57]. In this line, it should be noted that music is considered a kind of language, hence the interest in seeing if it reflects or shares neural resources with language. In this line, in an ERP work on musical perception [4] has been found that the processing of hierarchical structure with nested nonlocal dependencies -a mechanism fundamental for syntactic processing- is also activated during the perception of music. Therefore, it cannot yet be concluded that the musical syntactic process shares the bases of language but rather certain aspects. The different techniques inform us of various regions related to musical processing, although the exact differences in the syntactic treatment of language and music remain to be elucidated. In musicians' studies about musical phrasing, it is observed that the ERP shows a closure positive shift (CPS) in phrase boundaries -a positive shift in electrical activity at the closure of the phrase- [58–60] Also, the music CPS was observed in subjects of different cultural background listening both to music of their native and an alien culture. These findings add to the generality of the CPS as a marker for the processing of musical phrasing [61].

2.2 EEG channels interdependence measurements

The term functional connectivity (FC), is used to refer to the statistical interdependence between two neural signals (EEG or brain fMRI hemodynamic response signals) from anatomically different brain areas, a concept introduces by [62] and also defined as the temporal statistical correlation between spatially remote neurophysiological events between groups and dispersed neuronal areas [63]. Indeed, FC among different brain areas is important on brain processing since cognitive activity requires in general terms, that different brain regions not only co-act simultaneously, but there is also a functional interaction between them [64]. Furthermore, in the article by Núñez where FC was discussed in the human brain, it was reported that the cross interactions between local, regional and global networks are apparently responsible for a large part of the oscillatory EEG behavior [65]. In addition, this author report that combined EEG and high-resolution EEGs can provide different multiscale estimates of functional connectivity in healthy and diseased brains with measures such us covariance and coherence. In the field of musical perception, it has been reported that the analysis of the coherence or functional coupling between brain areas is of interest regarding the effect of music on the neurological mechanisms related to attention, cognition and emotion [66, 67].

2.2.1 EEG FC while listening to music

Authors have shown that musical perception requires the integration of different cortical areas [68]. This highly important concept has led researchers to use connectivity analysis [69–72] and network theory [73] to examine how different brain regions communicate while listening to music. In this line, related to musical experience, a study carried out in musicians shows synchronization of phase alterations in the alpha band between the right frontocentral cortical regions when musical expectation is violated [74]. Moreover, various EEG studies indicate that musical hearing produces changes in EEG coherence/synchronization in different bands [70, 74–77] and it has been considered of interest to study the configuration of the connectivity networks between different brain areas using modern graph theory that we will see later. Related to emotion, music-induced EEG neural correlations have been found at various frequencies on the prefrontal cortex and a set of functional connectivity patterns, defined by measures of coherence between channels, which are significantly different between the groups of emotional responses induced by music [78]. Recent studies show the integration of different cortical areas is required for musical perception and emotional processing [79] and that the magnitude of the cross-correlation values was significantly higher when we listened to unknown and coded music than when we listened to familiar music. These results are in agreement with those suggesting that the response to unfamiliar music is stronger than that of familiar music [80]. Furthermore, through joint EEG and fMRI spectral coherence measurements, a left cortical network has been identified that is involved with pleasant feelings associated with music [70]. These musical characteristics have been reported to produce greater sensory complexity of unexpected and puzzling situations or moments of unfulfilled expectations and higher levels of arousal [81].

2.2.2 Functional connectivity in musicians' experts

In general, it is known that the musician's brain has specific characteristics related to its functionality and structure [82–84]. There are numerous studies on this issue performed with signal analysis techniques (EEG and EMG) although most of them has been performed through BOLD neuroimagen fMRI signals. Thus, the size of the intranucal length of the precentral gyrus appears to be negatively correlated with age when their musical training starts has been found in keyboard players [85]. Furthermore, musically trained children show greater activation in areas related to executive functions like pre-supplementary motor area/supplementary motor area while performing a task [86]. In relation with FC in neuroimagen there is abundant literature carry out to expert musicians in a resting condition. Thus, musicians have been reported to exhibit stronger FC between the primary auditory cortex, the primary motor cortex [87] and in the right ventral premotor cortex. This is related to functional coupling between the motor and auditory areas and modulated as a function of musical training [88]. Also, in musicians, a significantly higher density of local functional connectivity has been shown in different brain regions [89], greater insular connectivity [90] and parietal opercular connectivity [91]. In musical interpretation condition, activity is reported in the auditory areas functionally connected with activity in the dorsal motor and pre-motor areas, whose connectivity is positively correlated with a good performance in interpretation [92]. Therefore, the musical experience seems to influence the functional connectivity (EEG) of some cortical areas. Indeed, in expert musicians, listening to extracts of tonal music modified the magnitudes of spectral coherence of the EEG in the alpha and beta bands with respect to non-musicians [93] and the phase synchronization of the gamma band, especially the left hemisphere [75]. Furthermore, the phase synchronization of expert musicians was greater than that of non-musicians listening to the same musical extract [76]. Another study reports that during the hearing of major and minor compositions by non-musicians, amateurs, and expert musicians, the EEG activity of the theta and gamma bands of the posterior cortical regions decreased with musical experience [94]. In this line, a study through the analysis of cortical images extracted from the ERPs and the responses of the subjects to the closure of complex musical stimuli (syntactic musical violations on which we have spoken in the univariate approach) reported important differences between groups, attributed to their different musical experience [95].

2.2.3 EEG during musical imagination in musicians

Finally, with regard to the work carried out professional musicians by analyzing signals electroencephalography (EEG) in Imagined interpretation is reported that this task proved to induce activation of the alpha band significantly stronger than the simple musical perception [96]. In this sense, it is known that musical images are a mental representation of music, as well as that its underlying mechanisms of perception are active and committed to it [97]. It is known that musical learning shows certain aspects of behavior that can be observed in the notable acquisition of skills of musicians, which is why the benefits of imagined interpretation in the learning of motor skills is a reason for interest and discussion among authors [98]. The imagined interpretation and the real interpretation or performance are correlated and are believed to activate similar neural structures [99]. There are some characteristics of the imagined interpretation, which we can call simulated action or mental rehearsal, that reveal a close relationship between it and motor action, specifically it has been pointed out that the synchronization patterns of both processes are similar [100] as well as that the changes in corticospinal excitability involve the same muscles in both conditions [101]. Consistent with this hypothesis, fMRI studies investigating imagined interpretation in paradigms where subjects execute hand and finger movements [102] have demonstrated activation of the supplemental motor area (SMA), the premotor cortex, the cerebellum and the primary motor cortex. Therefore, we can say that according to these studies, imagined performance and real performance share certain common characteristics reflected in the cerebral cortex and in the musculature. On the other hand, also in the fMRI technique, a study carried out on piano students [103] in which imagination and interpretation tasks were analyzed, found activations of the frontoparietal-bilateral network that includes the areas premotor, precuneus, and medial part of Brodmann's area (BA) 40 during both tasks. Other areas that appear to be involved in the imagined interpretation are the superior parietal and ventrolateral/dorsolateral frontal areas [104]. In another line of work, the activity of the EEG potentials was investigated in violinist players [105], finding that the bilateral frontal opercular regions are crucial both in the preparation and during the performance of music and during the imagination of the same in agreement with some previously commented fRMI results. The authors suggest that this effect is due to "mirror neurons" that are at the service of the observation or imagination of one's own performance [106]. It has been also observed the activation of different motor areas that were not the same for interpretation as for imagination. Functional interactions between the temporal cortex and the frontal cortex have been found to improve during musical imagination [107]. For all this, it seems that the imagined interpretation is capable of activating different areas of the cerebral cortex such as those belonging to the motor system, the SMA, the auditory cortex.

With the fMRI technique a study observes that, compared to the resting condition, the imagined interpretation increased in extended regions of the brain the FC of the supplementary motor area (SMA), including the sensorimotor cortices, the parietal cortex, the temporal cortex posterior cortex, occipital cortex and inferior and dorsolateral prefrontal cortex, this is related to cognitive control, motor planning and syntactic processing [108]. Increased connectivity with sensorimotor cortices is believed to be potentially involved with planning thought in motor programs. These authors also consider that the reconfiguration of the SMA network reflects the multimodal integration required for imagined musical interpretation and real interpretation/performance, as well as they propose that the SMA network build "the internal representation of musical performance" by integrating multimodal information required for the presentation [108]. The same authors, in a later study with the same task, found that imagined music performance increased the functional connectivity of the angular gyrus with different regions, which attributes a role to this region in the imagined performance [109]. Therefore, it is observed that the FC in the interpretation and imagined interpretation shares the configuration of networks that are involved in the performance process, which is different from neural activity, and therefore is able to connect in the imagined interpretation as in the real one. If the imagined interpretation is capable of activating the connections between brain regions that occur during interpretation, this is a way to study its possibilities from the perspective of imagination interpretation.

2.2.4 EEG FC using graph metric

In the graph theory context EEG channels are taken as the graph nodes and connectivity values between them as edges. The usefulness of this metric has been reported to be of interest in brain neural network research to evidence changes in its topological structure. Two measures are used to define different types of neural network organization: one involves the nodal groups, the clustering coefficient (C) and the other the magnitude of the length of the path between nodes, length of the characteristic path L. For a given node, C measures the tendency to link from neighboring nodes, reflecting the extent of the local domain; while L is associated with the ability to integrate global information and, therefore, with the readiness for communication within the brain [110, 111]. Depending on the relative magnitudes of C and L, different levels of topological organization of a cortical brain network are defined. Thus, a network is considered "regular" when a high value of C and L is obtained from its graph representation, while a network is considered "random" when a low value of C and L is found. Between both types of network, the type called small world (SW) is defined when a graph has a high C magnitude and a low L magnitude. Consequently, SW neural networks are said to have a high level of local information distribution together with a high efficiency for global transfer information, both properties of great relevance for the dynamics of complex brain processing [110, 111]. For determining the SW level of a network NN, the C and L magnitudes are normalized with regard to the mean of a number ($N = 100$) of random networks having the same number of nodes, edges, and degree distributions as the network NN [112]. A network with approximately equal L and larger C than matched random networks (i.e., normalized $L \sim 1$ and normalized $C > 1$) is said to be a SW network. In the context of the musical perception, listening to Chinese music (Guquin music excerpts versus silence and noise) in non-specialist subjects has been reported to produce an increase in functional connectivity (EEG phase coherence) in the alpha band, an improvement in cortical network organization of small world [73] and also a tendency to the random organization of the network as well -when a phase delay index is used that indicates a tendency to a more efficient but less economical architecture during musical listening [113]. Therefore, musical hearing somehow affects the topological structure of brain networks.

3. Conclusions

Since computational algorithms for signal analysis introduction in Biomedicine, different methods of cortical electrical signal analysis (EEG) have been used to study the neural multiple processes involved in musical perception. Applications range since from music recognition and its brain processing to its cognitive and emotional effects. In this broad chain of neural events, many brain centers and functions (central and peripheral) intervene. The participation and importance of some of these, by using different techniques of analysis and processing of EEG signals (including MEG and ERP cortical recordings) have been investigated along the time. In the review, the most interesting results appeared in the literature on the subject have been reported. Among them are those that study aspects such as musical syntax (its comparison with language), the differences between styles including consonances and dissonances, musical expectation and the nature of the different emotions (including rewards) produced by music. From the review carried out it is concluded that the analysis of cortical electrical signals (EEG, MEG, ERP) constitutes, mainly due to its high temporal resolution, a useful methodology for the study of many issues concerning the music-brain interaction.

Conflict of interest

The authors declare no conflict of interest.

Author details

González Almudena[1*], Santapau Manuel[2] and González Julián Jesús[3]

1 Art History Department, La Laguna University, Spain

2 String Department, Requena Conservatory, Spain

3 Department of Basic Medical Sciences (Physiology), La Laguna University, Spain

*Address all correspondence to: agonzabr@ull.edu.es

References

[1] Maess B, Koelsch S, Gunter T & Friederici A. (2001). Musical syntax is processed in Broca's area: A MEG study. Nature neuroscience. 4(5):540-5. doi: 10.1038/87502.

[2] Scherer K. (2004). Which Emotions Can be Induced by Music? What Are the Underlying Mechanisms? And How Can We Measure Them? Journal of New Music Research, 33: 239-251. doi:10.1080/0929821042000317822.

[3] Juslin PN & Västfjäll D. (2008). Emotional responses to music: the need to consider underlying mechanisms. Behav Brain Sci., 31(5):559-621. doi: 10.1017/S0140525X08005293.

[4] Koelsch S, Skouras S & Jentschke S. (2013). Neural Correlates of Emotional Personality: A Structural and Functional Magnetic Resonance Imaging Study. PLOS ONE 8(11): e77196. https://doi.org/10.1371/journal.pone.0077196.

[5] Pereira CS, Teixeira J, Figueiredo P, Xavier J, Castro SL, et al. (2011). Music and Emotions in the Brain: Familiarity Matters. PLOS ONE, 6(11): e27241.doi: 10.1371/journal.pone.0027241.

[6] Krumhansl CL & Cuddy LL. (2010). A theory of tonal hierarchies in music. In M. R. Jones (Ed.), Springer Handbook of Auditory Research. New York: Springer.

[7] Aube W, Peretz I & Armony J. (2013). The effects of emotion on memory for music and vocalisations. Memory (Hove, England). 21. doi: 10.1080/09658211.2013.770871.

[8] Collins T, Tillmann B, Barrett FS, Delbé C & Janata P. (2014). A combined model of sensory and cognitive representations underlying tonal expectations in music: from audio signals to behavior. Psychol Rev.,121(1):33-65. doi: 10.1037/a0034695.

[9] Bigand E, Tillmann B, & Poulin-Charronnat B. (2006). A module for syntactic processing in music? Trends in Cognitive Sciences, 10(5), 195-196. https://doi.org/10.1016/j.tics.2006.03.008.

[10] McDermott JH, Schultz AF, Undurraga EA, & Godoy RA. (2016). Indifference to dissonance in native Amazonians reveals cultural variation in music perception. Nature, 535, 547-550. doi: 10.1038/nature18635.

[11] Bowling DL, Hoeschele M, Gill KZ, W. & Fitch WT. (2017). The Nature and Nurture of Musical Consonance. Music Perception 1, 35(1): 118-121. doi:10.1525/mp.2017.35.1.118

[12] Rohrmeier MA & Koelsch S. (2012). Predictive information processing in music cognition. A critical review. Int J Psychophysiol., 83(2):164-75. doi: 10.1016/j.ijpsycho.2011.12.010. E

[13] Bailes F, Dean R & Pearce M. (2013). Music Cognition as Mental Time Travel. Scientific reports. 3. 2690. 10.1038/srep02690.

[14] Pearce M & Wiggins. (2012). Auditory Expectation: The Information Dynamics of Music Perception and Cognition. Topics in cognitive science. 4. 625-52. 10.1111/j.1756-8765.2012.01214.x.

[15] Gebauer L, Kringelbach ML & Vuust, P. (2012). Ever-changing cycles of musical pleasure: The role of dopamine and anticipation. Psychomusicology: Music, Mind, and Brain, 22(2), 152-167. https://doi.org/10.1037/a0031126.

[16] Huron D. (2006). Sweet Anticipation: Music and the Psychology of Expectation. 10.7551/mitpress/6575.001.0001.

[17] Curtis M & Bharucha J. (2009). Memory and Musical Expectation for Tones in Cultural Context. Music Perception - MUSIC PERCEPT. 26. 365-375. 10.1525/mp.2009.26.4.365.

[18] Soley G & Hannon E. (2010). Infants Prefer the Musical Meter of Their Own Culture: A Cross-Cultural Comparison. Developmental psychology. 46. 286-92. 10.1037/a0017555.

[19] Zatorre RJ & Salimpoor VN. (2013). From perception to pleasure: music and its neural substrates. Proceedings of the National Academy of Sciences of the United States of America, 110 Suppl 2(Suppl 2), 10430-10437. https://doi.org/10.1073/pnas.1301228110.

[20] Koelsch S. (2014). Brain correlates of music-evoked emotions. Nat Rev Neurosci 15, 170-180. https://doi.org/10.1038/nrn3666.

[21] Buzsáki G. (2006). Rhythms of the Brain. New York, NY: Oxford University Press.

[22] Koelsch S. (2011). Toward a Neural Basis of Music Perception – A Review and Updated Model. Frontiers in psychology. 2. 110. 10.3389/fpsyg.2011.00110.

[23] Bhattacharya J & Petsche H. (2005). Drawing on mind's canvas: Differences in cortical integration patterns between artists and non-artists. Human brain mapping. 26. 1-14. 10.1002/hbm.20104.

[24] Baumgartner T, Esslen M, & Jancke L. (2006). From emotion perception to emotion experience: emotions evoked by pictures and classical music. International Journal of Psychophysiology, 60, 34e43.

[25] Aftanas L, Reva N, Savotina L, & Makhnev V. (2006). Neurophysiological correlates of induced discrete emotions in humans: an individually oriented analysis. Neuroscience and Behavioral Physiology, 36, 119e130.

[26] Balconi M, & Lucchiari C. (2008). Consciousness and arousal effects on emotional face processing as revealed by brain oscillations. A gamma band analysis. International Journal of Psychophysiology, 67, 41e46.

[27] Tsang CD, Trainor LJ, Santesso DL, Tasker SL & Schmidt LA. (2001). Frontal EEG responses as a function of affective musical features. Ann N Y Acad Sci. 930:439-42. doi: 10.1111/j.1749-6632.2001.tb05764.x.

[28] Schmidt LA & Trainor LJ. (2001). Frontal brain electrical activity (EEG) distinguishes valence and intensity of musical emotions. Cognition and Emotion, 15(4): 487-500. https://doi.org/10.1080/02699930042000187.

[29] Altenmüller E, Schürmann K, Lim V & Parlitz D. (2002). Hits to the left, flops to the right: Different emotions during listening to music are reflected in cortical lateralisation patterns. Neuropsychologia, 40. 2242-56. doi: 10.1016/S0028-3932(02)00107-0.

[30] Trost W, Ethofer T, Zentner M & Vuilleumier P. (2012). Mapping aesthetic musical emotions in the brain. Cerebral Cortex, 22, 2769-83.

[31] Mitterschiffthaler M, Fu C, Dalton J & Andrew C & Williams S. (2007). Functional MRI study of happy and sad – affective states induced by classical music. Human brain mapping. 28. 1150-62. 10.1002/hbm.20337.

[32] Berridge KC, & Robinson TE. (2003). Parsing reward. Trends in Neurosciences, 26(9), 507-513. https://doi.org/10.1016/S0166-2236(03)00233-9.

[33] Chapin H, Jantzen K, Kelso JA, Steinberg F & Large E. (2010). Dynamic emotional and neural responses

to music depend on performance expression and listener experience. PLoS One, 5(12):e13812. doi: 10.1371/journal.pone.0013812.

[34] Blood AJ, Zatorre R, Bermudez P & Evans A. (1999). Emotional responses to pleasant and unpleasant music correlate with activity in paralimbic regions. Nature neuroscience, 2. 382-7. 10.1038/7299.

[35] Phan KL., Wager, T., Taylor, S. & Liberzon, I. (2002). Functional Neuroanatomy of Emotion: A Meta-Analysis of Emotion Activation Studies in PET and fMRI. NeuroImage. 16. 331-48.

[36] Brown, S., Martinez, M. J., & Parsons, L. M. (2004). Passive music listening spontaneously engages limbic and paralimbic systems. NeuroReport, 15(13), 2033-2037. http://dx.doi.org/10.1097/00001756-200409150-00008.

[37] Ramos J., y Corsi-Cabrera, M. (1989) Does brain electrical activity react to music? International Journal of Neuroscience, 47, 351-357.

[38] Sammler D., Grigutsch, M., Fritz, T. & Koelsch, S. (2007). Music and emotion: Electrophysiological correlates of the processing of pleasant and unpleasant music. Psychophysiology. 44. 293-304. 10.1111/j.1469-8986.2007.00497.

[39] Davidson JW. (1995) What Does the Visual Information Contained in Music Performances Offer the Observer? Some Preliminary Thoughts. In: Steinberg R. (eds) Music and the Mind Machine. Springer, Berlin, Heidelberg. https://doi.org/10.1007/978-3-642-79327-1_11

[40] Petrantonakis PC & Hadjileontiadis LJ. (2011). A novel emotion elicitation index using frontal brain asymmetry for enhanced EEG-based emotion recognition. IEEE Trans Inf Technol Biomed. 15(5):737-46. doi: 10.1109/TITB.2011.2157933.

[41] Lin YP., Wang, C.H., Jung, T.P., Wu, T.L., Jeng, S.K., Duann, J.R., et al. (2010). EEG- based emotion recognition in music listening. Biomedical Engineering, IEEE Transactions on, 57, 1798e1806.

[42] Breitling D., Guenther, W. & Rondot, P. (1987). Auditory perception of music measured by brain electrical activity mapping. Neuropsychologia, 25(5), 765-77. https://doi.org/10.1016/0028-3932(87)90114-X.

[43] Peretz I & Coltheart M. (2003). Modularity of music processing. Nature neuroscience. 6. 688-91. 10.1038/nn1083.

[44] Hadjidimitriou S & Hadjileontiadis L. (2013). EEG-Based Classification of Music Appraisal Responses Using Time-Frequency Analysis and Familiarity Ratings. IEEE Transactions on Affective Computing. 4. 10.1109/T-AFFC.2013.6.

[45] Wieser, H.G., & Mazzola, G. (1986). Musical consonances and dissonances: Are they distinguished independently by the right and left hippocampi? Neuropsychologia, 24(6), 805-812. https://doi.org/10.1016/0028-3932(86)90079-5

[46] Costa, M., Ricci Bitti, P. & Bonfiglioli, L. (2000). Psychological Connotations of Harmonic Musical Intervals. Psychology of Music - PSYCHOL MUSIC. 28. 4-22. 10.1177/0305735600281002

[47] Schön, D., Gordon, R. & Besson, M. (2006). Musical and Linguistic Processing in Song Perception. Annals of the New York Academy of Sciences. 1060. 71-81. 10.1196/annals.1360.006.

[48] Pallesen, K.J., Bailey, C.J., Brattico, E., Gjedde, A., Palva, J.M, et al. (2015)

Experience Drives Synchronization: The phase and Amplitude Dynamics of Neural Oscillations to Musical Chords Are Differentially Modulated by Musical Expertise. PLOS ONE 10(8): e0134211. https://doi.org/10.1371/journal.pone.0134211

[49] Babiloni, C., Buffo, P., Vecchio, F., Marzano, N., Del Percio, C., Spada, D., Rossi, S., Bruni. I., Rossini, P.M. & Perani, D. (2012). Brains "in concert": frontal oscillatory alpha rhythms and empathy in professional musicians. Neuroimage, 60(1):105-16. doi:10.1016/j.neuroimage.2011.12.008.

[50] Patel, A., Gibson, E., Ratner, J., Besson, M. & Holcomb, P. (1998). Processing Syntactic Relations in Language and Music: An Event-Related Potential Study. Journal of cognitive neuroscience. 10. 717-33. 10.1162/089892998563121.

[51] Patel, A. (2003). Patel, A. D. Language, music, syntax and the brain. Nature Neurosci. 6, 674-681. Nature neuroscience. 6. 674-81. 10.1038/nn1082.

[52] Miranda, R.A., & Ullman, M. T. (2007). Double dissociation between rules and memory in music: an event-related potential study. Neuroimage, 38, 331-345.

[53] Koelsch, S. (2009). Universal recognition of three basic emotions in music. Current Biology, 19(7), 573-576.

[54] Koelsch, S., & Mulder, J. (2002). Electric brain responses to inappropriate harmonies during listening to expressive music. Clinical Neurophysiology, 113(6), 862-869. https://doi.org/10.1016/S1388-2457(02)00050-0

[55] Koelsch S & Jentschke S. (2010). Differences in electric brain responses to melodies and chords. J. Cogn. Neurosci. 22, 2251-2262.

[56] Koelsch S, Gunter T, Wittfoth M & Sammler D. (2005b). Interaction between syntax processing in language and in music: an ERP study. J. Cogn. Neurosci. 17, 1565-1577.

[57] Steinbeis N & Koelsch S. (2011). Affective priming effects of musical sounds on the processing of word meaning. J. Cogn. Neurosci. 23, 604-621.

[58] Knösche T, Neuhaus C, Haueisen J, Alter K, Maess B, Witte O & Friederici AD. (2005). Perception of phrase structure in music. Hum. Brain Mapp. 24, 259-27310.1002/hbm.20088.

[59] Neuhaus C, Knösche T & Friederici A. (2006). Effects of musical expertise and boundary markers on phrase perception in music. J. Cogn. Neurosci. 18, 472-493.

[60] Peter, V., McArthur, G. & Crain, S. (2014). Using event-related potentials to measure phrase boundary perception in English. BMC neuroscience. 15. 129. 10.1186/s12868-014-0129-z.

[61] Nan, Y., Knösche, T. & Friederici, A. (2006). The perception of musical phrase structure: a cross-cultural ERP study. Brain Res. 1094, 179-19110.1016/j.brainres.2006.03.115.

[62] Friston, K.J., Frith, C.D. & Frackowiak, R.S.J. (1993), Time-dependent changes in effective connectivity measured with PET. Hum. Brain Mapp., 1: 69-79. doi:10.1002/hbm.460010108.

[63] Lang, E.W., Tomé, A.M., Keck, I.R., Górriz-Sáez, J.M. & Puntonet., C.G. (2012). Brain Connectivity Analysis: A Short Survey. Comput Intell Neurosci., 412512. doi: 10.1155/2012/41251.

[64] Friston, K.J. and Price, C.J. (2001), Generative models, brain function and neuroimaging. Scandinavian

Journal of Psychology, 42: 167-177. doi:10.1111/1467-9450.00228.

[65] Nuñez P.L, Srinivasan R. & Fields RD. (2015). EEG functional connectivity, axon delays and white matter disease. Clin Neurophysiol., 126(1):110-20. doi: 10.1016/j.clinph.2014.04.003.

[66] Nuñez P.L., Wingeier B.M., Silberstein R.B. (2001). Spatial-temporal Structures of Human Alpha Rhythms: Theory, Microcurrent Scources, Multiscale Measurements, and Global Binding of Local Networks. Human Brain Mapping, 13: 125-164.

[67] Singer W. (1999). Neuronal synchrony: a versatile code for the definition of relations? Neuron 24 49-65, 111-125. 10.1016/s0896-6273(00)80821-1.

[68] Miskovic, V. & Schmidt, L. (2010). Cross-regional cortical synchronization during affective image viewing. Brain research. 1362. 102-11. 10.1016/j.brainres.2010.09.102.

[69] Bhattacharya, J., & Petsche, H. (2005). Phase synchrony analysis of EEG during music perception reveals changes in functional connectivity due to musical expertise. Signal Processing, 85, 2161e2177.

[70] Flores-Gutiérrez, Enrique & Díaz, José-Luis & Barrios, Fernando & Favila-Humara, Rafael & Guevara, Miguelangel & Río-Portilla, Yolanda & Corsi-Cabrera, María. (2007). Metabolic and electric brain patterns during pleasant and unpleasant emotions induced by music masterpieces. International journal of psychophysiology: official journal of the International Organization of Psychophysiology. 65. 69-84. 10.1016/j.ijpsycho.2007.03.004.

[71] Karmonik, C., Brandt, A., Fung, S., Grossman, R. & Frazier, J. (2013). Graph theoretical connectivity analysis of the human brain while listening to music with emotional attachment: Feasibility study. Conference proceedings: Annual International Conference of the IEEE Engineering in Medicine and Biology Society. IEEE Engineering in Medicine and Biology Society. Co. 6526-6529. 10.1109/EMBC.2013.6611050.

[72] Kay, B.P., Meng, X., Difrancesco, M.W., Holland, S.K. & Szaflarski, J.P. (2012). Moderating effects of music on resting state networks. Brain Res.4, 1447: 53-64.

[73] Wu, J., Zhang, J., Liu, C., Liu, D., Ding, X., & Zhou, C. (2012). Graph theoretical analysis of EEG functional connectivity during music perception. Brain Research, 1483, 71e81.

[74] Ruiz M. H., Jabusch H.C., Altenmüller E. (2009). Detecting wrong notes in advance: neuronal correlates of error monitoring in pianists. Cereb. Cortex, 19: 2625-2639 10.1093/cercor/bhp021.

[75] Bhattacharya, J. & Petsche, H. (2001a). Universality in the brain while listening to music. Universality in the brain while listening to musicProc. R. Soc. Lond. B.2682423-2433. https://doi.org/10.1098/rspb.2001.1802

[76] Bhattacharya, J., Petsche, H., Pereda, E. (2001b). Long-range synchrony in the gamma band: role in music perception. Journal of Neuroscience, 21 6329-6337.

[77] Petsche, H., Kaplan, S., Von Stein, A. & Filz, O. (1997). The possible meaning of the upper and lower alpha frequency ranges for cognitive and creative tasks. International Journal of Psychophysiology, 26, (1-3): 77-97.

[78] Daly I, Malik A, Hwang F, Roesch E, Weaver J, Kirke A,

Williams D, Miranda E, Nasuto SJ. (2014). Neural correlates of emotional responses to music: an EEG study. Neurosci Lett., 573:52-7. doi: 10.1016/j.neulet.2014.05.003.

[79] Shahabi, H. & Moghimi, S. (2016). Toward automatic detection of brain responses to emotional music through analysis of EEG effective connectivity. Computers in Human Behavior. 58. 231-239. 10.1016/j.chb.2016.01.005.

[80] Kumagai Y., Arvaneh, M., & Tanaka, T. (2017). Familiarity Affects Entrainment of EEG in Music Listening. Frontiers in human neuroscience, 11, 384. https://doi.org/10.3389/fnhum.2017.00384

[81] Bodner E, Gilboa A & Amir DD. (2007). The unexpected side-effects of dissonance. Psychology of Music - PSYCHOL MUSIC. 35. 286-305. 10.1177/0305735607070381.

[82] Schlaug G. (2001). The brain of musicians: A model for functional and structural adaptation. Annals of the New York Academy of Sciences. 930. 281-99.

[83] Zatorre R., Chen, J. & Penhune, V. (2007). When the brain plays music: Auditory-motor interactions in music perception and production. Nature reviews. Neuroscience 8: 547-58. 10.1038/nrn2152.

[84] Sittiprapaporn W. (2012). The Musician's Brain. Journal of Biological Sciences, 12: 367-375.

[85] Amunts K., Schlaug, G., Jäncke, L., Steinmetz, H., Schleicher, A., Dabringhaus, A. & Zilles, K. (1997). Motor cortex and hand motor skills: Structural compliance in the human brain. Human Brain Mapping, 5, 206-215. doi:10.1002/(SICI)1097-0193(1997)5:3<206:AID-HBM5>3.0.CO;2-7.

[86] Zuk J., Benjamin, C., Kenyon, A. & Gaab, N. (2014). Behavioral and neural correlates of executive functioning in musicians and non-musicians. PLoS One. 2014 Jun 17;9(6):e99868. doi: 10.1371/journal.pone.0099868. Erratum in: PLoS One. 2015;10(9):e0137930. Erratum in: PLoS One. 2018 Jan 11;13(1):e0191394. PMID: 24937544; PMCID: PMC4061064.

[87] Luo C., Guo, Z., Lai, Y., Liao, W., Liu, Q., et al. (2012) Musical Training Induces Functional Plasticity in Perceptual and Motor Networks: Insights from Resting-State fMRI. PLOS ONE 7(5): e36568. https://doi.org/10.1371/journal.pone.0036568.

[88] Palomar-García MA, Zatorre, R.J., Ventura-Campos, N., Bueichekú, E. & Ávila. C. (2017). Modulation of Functional Connectivity in Auditory-Motor Networks in Musicians Compared with Nonmusicians. Cereb Cortex, 27(5):2768-2778. doi: 10.1093/cercor/bhw120.

[89] Luo C., Tu, S., Peng, Y., Gao, S., Li, J., Dong, L., Li, G., Lai, Y., Li, H., & Yao, D. (2014). Long-term effects of musical training and functional plasticity in salience system. Neural plasticity, 2014, 180138. https://doi.org/10.1155/2014/180138.

[90] Zamorano AM, Cifre, I., Montoya, P., Riquelme, I. and Kleber, B. (2017), Insula-based networks in professional musicians: Evidence for increased functional connectivity during resting state fMRI. Hum. Brain Mapp., 38: 4834-4849. doi:10.1002/hbm.23682.

[91] Tanaka S & Kirino E. (2017b). The parietal opercular auditory-sensorimotor network in musicians: A resting-state fMRI study. Brain and Cognition. 120. 10.1016/j.bandc.2017.11.001.

[92] Segado M., Hollinger, A., Thibodeau, J., Penhune, V., & Zatorre,

R. J. (2018). Partially Overlapping Brain Networks for Singing and Cello Playing. Frontiers in neuroscience, 12, 351. https://doi.org/10.3389/fnins.2018.00351

[93] Petsche H., Richter, P., Von Stein, A., Etlinger, S., & Filz, O. (1993). EEG Coherence and Musical Thinking. Music Perception: An Interdisciplinary Journal, 11(2), 117-151. doi:10.2307/40285613.

[94] Jenni R., Oechslin, M. & James, C. (2017). Impact of major and minor mode on EEG frequency range activities of music processing as a function of expertise. Neuroscience Letters. 647. 10.1016/j.neulet.2017.03.022.

[95] James C.E., Cereghetti ,D.M., Roullet Tribes, E. & Oechslin M.S. (2015). Electrophysiological evidence for a specific neural correlate of musical violation expectation in primary-school children. Neuroimage, 104, 386-397. 10.1016/j.neuroimage.2014.09.047.

[96] Schaefer RS., Morcom, A.M., Robert,s N. & Overy, K. (2014). Moving to music: effects of heard and imagined musical cues on movement-related brain activity. Front Hum Neurosci. 26(8):774. doi: 10.3389/fnhum.2014.00774. PMID: 25309407; PMCID: PMC4176038.

[97] Schaefer RS., Vlek R. J., Desain P. (2011b). Music perception and imagery in EEG: alpha band effects of task and stimulus. Int. J. Psychophysiol. 82, 254-25910.1016/j.ijpsycho.2011.09.007.

[98] Langheim, F.J., Callicott, J.H., Mattay,V.S., Duyn, J.H. & Weinberger, D.R. (2002). Cortical Systems Associated with Covert Music Rehearsal. Neuroimage, 16, 901-908.

[99] Jeannerod M. (1995). Mental imagery in the motor context. Neuropsychologia 33, 1419-1432 10.1016/0028-3932(95)00073-C.

[100] Sirigu A & Duhamel JR. (2001). Motor and visual imagery as two complementary but neurally dissociable mental processes. J Cogn Neurosci., 13(7):910-9. doi: 10.1162/089892901753165827.

[101] Fadiga L, Buccino, G., Craighero, L., Fogassi L, Gallese, V. & Pavesi, G. (1999). Corticospinal excitabili,ty is specifically modulated by motor imagery: a magnetic stimulation study. Neuropsychologia, 37(2):147-58. doi: 10.1016/s0028-3932(98)00089-x.

[102] Lotze M., Montoya P, Erb, M., Hülsmann, E., Flor, H., Klose, U., Birbaumer N & Grodd W. (1999). Activation of cortical and cerebellar motor areas during executed and imagined hand movements: an fMRI study. J Cogn Neurosci. 11(5):491-501. doi: 10.1162/089892999563553.

[103] Meister I.G., Krings T, Foltys H, Boroojerdi B, Müller M, Töpper R & Thron, A. (2004). Playing piano in the mind--an fMRI study on music imagery and performance in pianists. Brain Res Cogn Brain Res., 19(3):219-28. doi: 10.1016/j.cogbrainres.2003.12.005.

[104] Zatorre R, Halpern, A. & Bouffard, M. (2009). Mental Reversal of Imagined Melodies: A Role for the Posterior Parietal Cortex. Journal of cognitive neuroscience. 22. 775-89. 10.1162/jocn.2009.21239.

[105] Kristeva R, Chakarov, V., Schulte-Mönting, J. & Spreer, J.(2003). Activation of cortical areas in music execution and imagining: a high-resolution EEG study. Neuroimage, 20(3):1872-83. doi: 10.1016/s1053-8119(03)00422-1.

[106] Nirkko A, Baader A, Lövblad KO, Milani P & Wiesendanger M. (2000). Cortical representation of music production in violin players: Behavioral assessment and functional imaging of finger sequencing, bimanual

coordination and music specific brain activation. Neuroimage, 11. 10.1016/S1053-8119(00)91039-5.

[107] Herholz SC, Halpern AR & Zatorre RJ. (2012). Neuronal correlates of perception, imagery, and memory for familiar tunes. J Cogn Neurosci., 24(6):1382-97. doi: 10.1162/jocn_a_00216.

[108] Tanaka S & Kirino E. (2017). Dynamic reconfiguration of the supplementary motor area network during imagined music performance. Front. Hum. Neurosci. 11:606. doi: 10.3389/fnhum.2017.00606.

[109] Tanaka S & Kirino E. (2019). Increased Functional Connectivity of the Angular Gyrus During Imagined Music Performance. Frontiers in Human Neuroscience, 13, 2. Doi: 10.3389/fnhum.2019.00092.

[110] Rubinov M & Sporns O. (2010). Complex network measures of brain connectivity: uses and interpretations. Neuroimage, 52(3):1059-69. doi: 10.1016/j.neuroimage.2009.10.003.

[111] Watts D & Strogatz S. Collective dynamics of 'small-world' networks. Nature 393, 440-442 (1998). https://doi.org/10.1038/30918

[112] Humphries MD & Gurney K. (2008). "Network 'small-world-ness': a quantitative method for determining canonical network equivalence". PLOS ONE. 3 (4): e0002051. Bibcode:2008PLoSO...3.2051H. doi:10.1371/journal.pone.0002051.

[113] Wu D, Li C & Yao D. (2013). Scale-free brain quartet: artistic filtering of multi-channel brainwave music. PLoS ONE 8:e64046. doi: 10.1371/journal.pone.0064046.

Chapter 3

Multicriteria Algorithm for Multisensory Food Analysis

Alexandre Magno da Rocha Vianna

Abstract

Observation of the multisensory experience using fMRI, EEG and analysis of users' responses using Fuzzy logic. The tabulation of these data aims to verify the responses and quantify them for comparison with a personal opinion survey using the SAATY scale and apply in general terms (opinion response and neural response) in other users belonging to the same group of people. Based on the data and processes of the described applications, a sensory mapping of the observer and verification of "patterns in neurophysiological processes" can be performed with a verbal response, in addition to allowing the understanding of the importance of these patterns for the selection of an option.

Keywords: multisensory, neurodesign, neuroscience, food

1. Introduction

This research is a continuous process of studies started in 2016 at UFRJ under the guidance of Dr. Paulo de Oliveira Reis through LabFuzzy. The proposal deals with the use of neuroscientific technologies to analyze and identify multisensory analogues that can improve the users' experience in the design process. An analysis of how human cognitive processes (related to the acquisition of information such as: learning, attention, memory, language, reasoning, decision making, etc.) linked to people's behavior has been analyzed and applied for some decades. A good example is given by Simon [1], when studying the human cognitive processes linked to decision making. This field was developed and branched out into the decision process linked to the marketing processes.

Subsequently, psychologists Khaneman and Tversky [2, 3] developed the Prospect Theory, demonstrating a profit and loss perspective model and the cognitive impact related to this event, which demonstrated that people tend to value much more avoiding losses than generating gains. An important milestone along this path occurred in 1980, when the first full-body magnetic resonance imaging (MRI) occurred in Scotland, a major advance compared to X-rays. In the middle of the following decade, functional magnetic resonance imaging (fMRI) recorded, for the first time, images of the brain in activity, during and after stimuli, showing the activation of the regions involved in certain tasks without the need to use ionizing radiation and exogenous contrast. Like traditional MRI devices. This model brings a new perspective in the view of consumption that, from the 90's, is widely treated in marketing processes. Lindstrom [4] through research on thousands of brains, analyzed a large part of the processes related to attractiveness and consumption behavior.

This chapter proposes the use of multidisciplinary concepts and tools between Creative Economy, Neuroscience, Mathematics and design for the observation of multisensory experience. The author of this chapter [5] enters this universe through the Design of furniture for offshore ships, where in addition to the functional character the internal environments need to attend to the psychological of the employees who remain there for months. Colors, shapes, spaces, art, everything has to be thought with a view to physical and emotional health. From that moment on (2006), a methodology for designing products and images with a focus on the user experience starts to be constructed through multisensory analogs, that is, information sensorialized by the observers that arouse certain reactions and that can be applied to the 5 senses. This methodology was used in the industrial sector and in micro and small companies (healthy food, PET, drinks, food, etc.) and patterns that can be recognized and mapped to generate certain reactions in the consumer were observed. As of 2017, with the acquisition of fMRI and EEG software and equipment, the Fuzzy Logic started to be used to quantify the mapped responses that started to be studied and tested in applications related to food and beverages.

Presentation of the techniques used for research in this chapter:

- Functional magnetic resonance (fMRI) analysis mentioned above,

- Electroencephalogram (EEG): used to record the electrical activity of the brain that is diagnosed by means of neural oscillations or brain waves, beta, alpha, theta and zeta.

- Hormonal dosage of saliva: just as the blood test can identify hormones and other elements, the analysis of saliva focuses on those present at the time of collection).

Other techniques can be used in data collection such as: Galvanic skin conductance (SCR), Skin temperature analysis by infrared thermography (ING), Heart rate (HVR), Facial expression recognition (FER), Implicit association test (IAT), retina mapping (eye tracking) and other less common.

For the research to be presented, techniques such as functional magnetic resonance analysis (fMRI), electroencephalogram (EEG), hormonal dosage of saliva were used, in addition to other techniques such as focus group and SAATY scale. The objective is to search the data for neurophysiological patterns that can generate a division between satisfactory and unsatisfactory neurobehaviors, that is, unfavorable reactions occur when data is presented and favorable reactions occur in the same way. The construction of these favorable or satisfactory reactions can be applied widely (a city, a public space, an enclosed space) or in very small areas (an environment, an object, a product, a service experience or even physical) or digital image). The point in question is the result that must be designed so that the user has an intended experience and keeps it in his memory. This analysis will be done using the techniques obtained above to identify in users, areas of the brain that have been stimulated and to identify these patterns, to be determined in each element of the project.

To understand what this observer seeks, it is necessary to go through the motivation theorists: Freud [6], Maslow [7] and Herzberg [8] who explain the subjectivity of human thought and how it correlates with social cycles and their own experience with what is observed, generating "satisfactory and unsatisfactory" profiles, rationalizing their environment and their sensorializations. Freud develops his thinking about the subjectivity of human choices, through his literatures: The unconscious, which traces a trajectory of ideas of the subjective factors of our choices. This thought was addressed with the example of "motivation for choosing an object" that occurs due to factors such as color, weight, size, texture, shape, etc. which are linked to the interpretation of each of these details through subjective memory, as well as the "object as a whole", as well as, all "emotional charge" contained in its interpretation in our past and its relationship with the present moment.

Maslow in his work "Motivation and Personality" describes the "Hierarchy of needs" which is widely known as the "Maslow's Pyramid". According to him, the individual feels the desire to satisfy his needs, according to the stage or level. Therefore, the motivation to fulfill these wishes comes gradually, generating a certain predictability of needs and desires according to several sociodemographic (age, income, geographic area, etc.) and behavioral (ambitious, heavy users, etc.) factors as well as in the cycle of life of a person or group of individuals. Herzberg, talks about the "2-factor theory" that defends the idea that there are satisfactory and unsatisfactory motivators in the work environment and that it is used in marketing management Kotler [3] for the marketing universe interpreting these concepts for products, services, people, environments, etc. From there, it is necessary to understand what makes the observer have a pleasant experience and enter the universe of experience design. The one by Csikszentmihalyi [9] brings in his works about Flow (concept defined as the great experience, or the psychology of the great experience, is linked to the challenge and reward zones of the brain and related to learning) the aspects that are related to the experience engaging and charming enough to pass the time and even forget some of the most basic needs like eating. To reach this level of immersion, Norman's work [10] and his concepts on design bring terms to specify levels of experience of the observer that go beyond the visceral or subconscious level, to the reflective level, but this reflection is based on their habits, which in turn, is at the behavioral and instinctual level, which is basically the automatic and subjective area linked to our experience. An individual's experience contains memories and emotional charges linked to the result of similar previous situations and which, in turn, are the result of our emotions Krippendorff [11].

The projected emotions will be mainly in search of the observer's pleasure and satisfaction as ways to reach the state of Flow and in this context the pleasures of Jordan 4 [12] relate aspects of functionality and usability to these states. The author assumes that pleasures are the result of the hedonic and practical emotional benefits associated with a product and that needs related to the usability of that product can be satisfied in 4 forms of pleasure (physiological pleasure, social pleasure, psychological pleasure and ideological pleasure). These authors reflect the concepts and tools for the investigation and manifestation of multisensory experiences. Another literature such as Logic Fuzzy Tanaka [13] is needed to measure the observer's reaction to the possibilities. In Fuzzy Logic, a multiple-value chain can be generated in which the logical values of the variables can be any real number between 0, corresponding to the false value, and 1, corresponding to the true value, for example. In contrast, in Boolean logic, the logical values of variables can be just 0 and 1, a or b, etc. this means that there are several possibilities between true and false and these possibilities can be found and determined if they correspond to the analyzed objective with the greatest probability, possibility and or plausibility. In this way, measurable variables can be found and an analysis system designed to measure these experiences created Ross [14]. One should not only take into account the tabulation of these data, but obtain them to measure the reaction of the observer and verify through the emotional dimensions of Carter [15] the impact that will provide for them to reach a certain objective, be it to obtain better performances or simply for entertainment.

It works like this, people determine verbally when asked about a particular experience or random information, but the verbal response does not always represent the cognitive response. In addition to the truth, there are exaggerations, minimizations and lies. This happens for several reasons and more often than we imagine Feldman [16]. However, by analyzing the cognitive response by means of Functional Magnetic Resonance (fMRI), Electroencephalogram (EEG), hormonal dosage of saliva and other previously necessary techniques, it can be verified if certain areas of the brain have been affected and what is the biological response. For example, when smelling a food being made on the spot, the olfactory cortex in the temporal lobe will be stimulated and information will be sent to areas such as the frontal lobe cortex (association with similar previous experiences), limbic system of the diencephalon (Memory) and response through neural peptides available in the hippocampus Silverthorn [17]. Other areas can be stimulated and hormones such as ghrelin for example, also known as "hunger hormone", are a peptide produced mainly by epsilon cells in food and by the pancreas when on an empty stomach, which then acts on the lateral hypothalamus and in the arched nucleus, generating a sensation of hunger. These areas of the brain can be observed by means of magnetic resonance in conjunction with specific areas (nucleus accumbens, geniculate body and certain areas of the visual cortex) making it possible to observe the option for a particular dish in a restaurant, in addition to determining the intensity of these reactions through EEG (stress, engagement, interest, excitement, focus and relaxation). One can go further, and check the levels of Ghrelin, Serotonins and Dopamines (through the analysis of saliva), understanding the levels of hormones and peptides Dispenza [18], responsible for hunger, pleasure and well-being. It can be measured against a quantity when you see, taste, touch, smell, etc. in each of the restaurant's dish options, following the previous example. When determining the preference standards of the tested group (which should represent consumers and target customers through personas, for example), it is possible to select a better menu, presentation of dishes, images for campaigns, restaurant environment, employee uniform, memorabilia and etc.

The complexity of the new consumer market requires an investigative method that clearly demonstrates how to design experiences for a group, capable of generating a positive impact, focusing on the necessary elements and increasing performance, minimizing unnecessary and incorrectly applied costs. The application of an investigative method linked to the relationship between the public and a medium or product is related to the sociocultural profile of these people in the sense that their parameters of relationship with the external world and interpretations of this world come from the knowledge acquired throughout life, where a large part of their cognitive stimuli and co-relationships happen through the environment where they live and relate daily.

1.1 Stages of development

Phase 1. Qualitative Research: The generation of personas is necessary to define a group that can represent the general public.

Phase 2. Explanatory research (cause and effect): The processes of attention, retention and memory, occur in different areas of the brain and transmit information to the body through hormones and peptides that are used as a basis to measure whether the data obtained in the experiment is positive or negative (EEG, fMRI and saliva analysis).

Phase 3. Experimental Research. Observation of the user's brain during the experiment.

2. Experiences

The experiences presented below were evolutions and variations of this methodology in the area of food. Because it is a sector that can awaken the 5 sensory channels, even though the 3 main ones are vision, smell and taste. The first test was developed using the image of 3 foods (as we can see in **Figures 1–3**), where a more juicy image was presented (**Figures 1** and **2**) and even an image with desaturated and manipulated foods to look unpleasant (**Figure 3**). The objective of having 3 such different variations, was to verify if the neural and verbal response coincided in options that seemed to be practically unanimous and if the result of the algorithm would follow this path, **Figure 1** was selected as the most tasty and **Figure 3** the least tasty as we expected.

Subsequently, 3 hamburger options were tested (**Figures 4–6**) this time, the test was to understand whether the launch of a new hamburger option with reference to the artisanal hamburger trend in Brazil (**Figure 6**) would be accepted among the more consolidated options on the market (**Figures 4** and **5**). This time, the acceptance of the new option was rejected by the participants in the verbal analysis (**Figure 8**), but an interesting feature was observed in the EEG analysis (**Figure 7**). According to the software data, the level of "interest" in the participants increased when this option was presented, which led me to invite them to a focus group to try to understand this discrepancy between verbal and neural analysis. The response of the participants in most cases was that they had an interest in trying the option in

Figure 1.
Image of barbecue with grilled fruits.

Figure 2.
Pasta with fruits.

Figure 3.
Plate of fruits, fungi, sausage and bacon with sauce (with manipulation of the saturation of the colors to look disgusting).

Figure 4.
Hamburguer A.

Figure 5.
Hamburguer B.

Figure 6, but they were not willing to "pay" for that option, the options in **Figures 4** and **5** being better known, that is, even in no information has ever been given on option values, whether there would be any payment, etc., the participants' response had a direct connection with the fact that they preferred to stick to traditional options rather than try a new one (taking into account automatically) the financial factor and the fact that they prefer to stick to already known tasty options

Figure 6.
Hamburguer C.

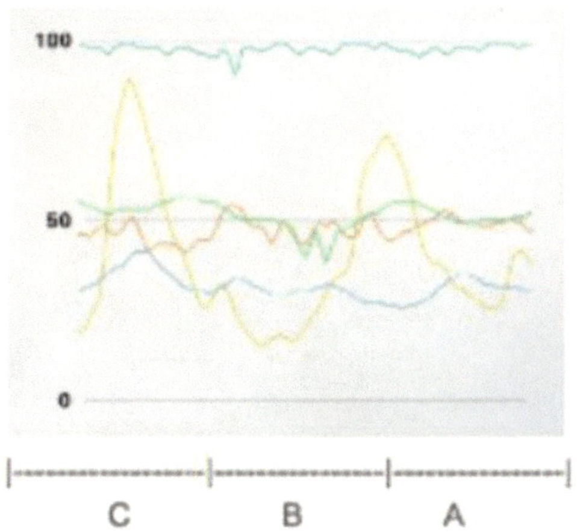

Figure 7.
EEG Interest is the yellow wave.

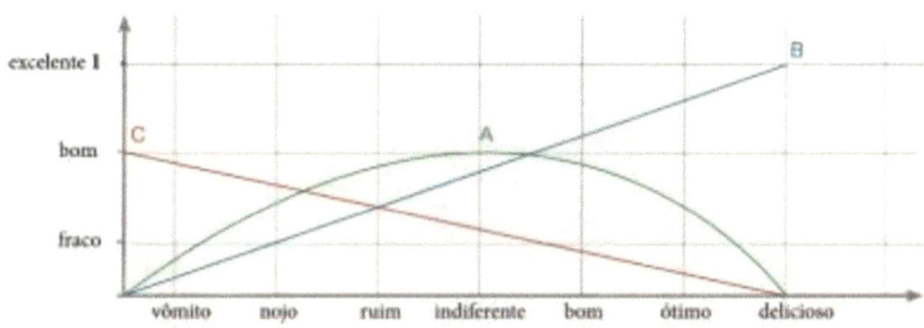

Figure 8.
Although the waves show high interest in options C and A, the verbal chart portrays them as the worst options.

than to take a chance on a new flavor that is likely not to be as tasty, according to Prospect theory).

The following analysis (**Figures 9–11**) was performed without the hormonal dosage of saliva, however, EEG analyzes were introduced in a more protagonist way. This decision was made in order to better observe the experience because it is understood that when it comes to food, when eating a salad or a chocolate, most of the stimulated areas observed through the fMRI equipment are the same. But the preference for the second option (chocolate) is probably greater in most people,

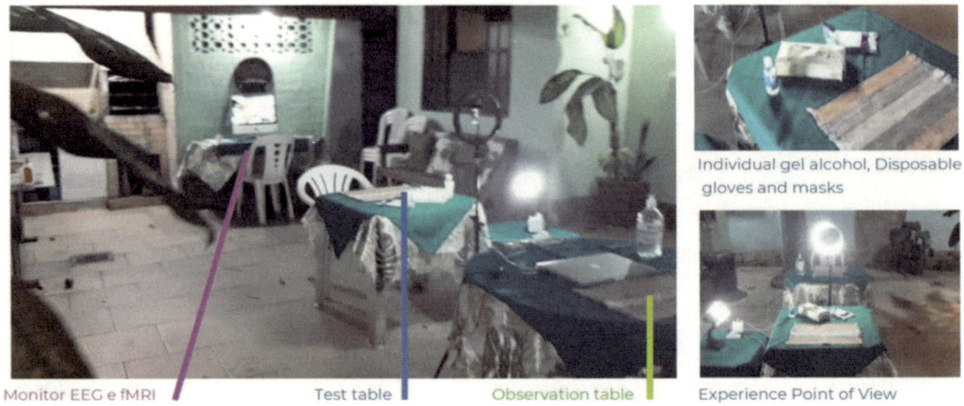

Figure 9.
Analysis site of coffee samples.

Figure 10.
Sample A.

Figure 11.
Sample B.

simply because the brain prefers more fatty foods due to its high energy consumption (it consumes approximately 20% of the energy produced by the body). Therefore, the analysis of waves and neural frequencies, as well as the attributes presented by the EEG software, are faster to interpret. In addition to this facilitation, it is possible to compare the results verified by the fMRI images (brain waves are represented in different colors, with 1 color for each frequency - **Figure 15**). Another way of comparing the 3 techniques presented (fMRI, EEG and hormonal dosage of saliva) is precisely through the hormones found during the experiments, as well as their quantity. The hormones, wave frequencies and analysis of the EEG software, demonstrate differently the possible emotional states of the participants and can be compared with each other.

2.1 Analyzed areas

Lateral geniculate body and prefrontal cortex: attraction for appearance. The appearance of food is more important than taste in many cases, even when food is prepared with the same ingredients and cooking methods. When we see something we shouldn't be eating, the brain sends an alarm. You can be fooled by stimuli of the smell of good food and the sound of frying, for example, which stimulate the production of ghrelin.

Accumbens Core: Food generates dopamines and serotonins and the tastier the better (our brain tends to prefer foods with more fat). It is related to pleasure.

Olfactory cortex: smell is one of the oldest senses and its stimulus is received in the olfactory bulb that directs the stimulus to the olfactory cortex, limbic system, motor cortex and others.

Visual cortex: related memory and visual processing.

Prefrontal cortex: related to planning complex behaviors and thoughts, expression of personality, decision making and modulation of social behavior.

2.2 Personas selection

In the case of a complex analysis, the cut used so far is that of qualitative research, seeking to analyze a series of people who can represent the target audience with quality. An adopted methodology uses a method of crossing supply and demand matrices Cosenza [19] to identify the distances between the different product alternatives in meeting the needs of its customers according to the profile demanded by them.

2.3 Phase identification (coffee experience)

- Personas. Selection.
- People who represent personas (**Figures 9–18**).
- Subscription.
- Sensory scale (linguistic variables and neural response).
- Multicriteria algorithm for sensory analysis. fMRI and EEG. SAATY scale for verbal response (1, 2, 3, 4, 5, 7, 9).
- Observation of the impacted area.
- Fuzzification and defuzzification of results (normative maximization).
- Answer 1: Selection of the option.
- Answer 2: evaluation of the experience at a rational and emotional level.

Figure 12.
Sample C.

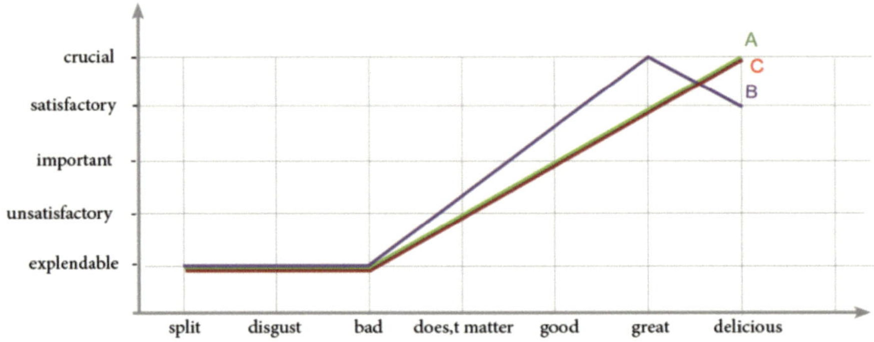

Figure 13.
Verbal Response graph, Sample A, B and C.

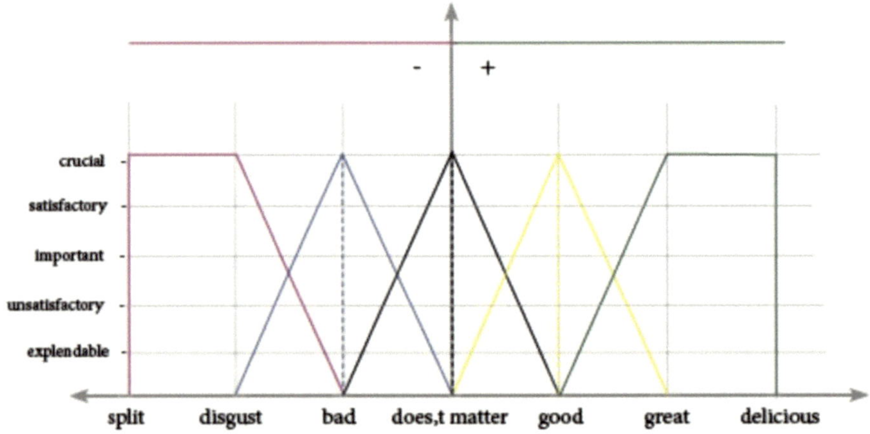

Figure 14.
Verbal Attributes description fuzzy matrix.

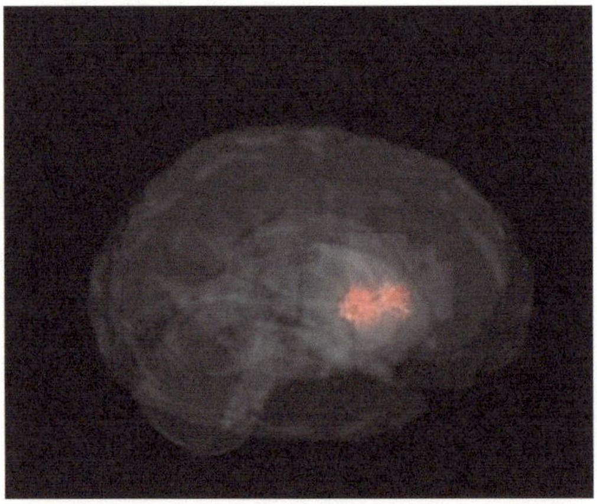

Figure 15.
fMRI.

Step 1: Preparation of the drink. Sensory analysis was performed using 3 coffee samples. For the preparation, a conventional home coffee maker was used. The coffee samples were prepared using coffee and mineral water. Each infusion was

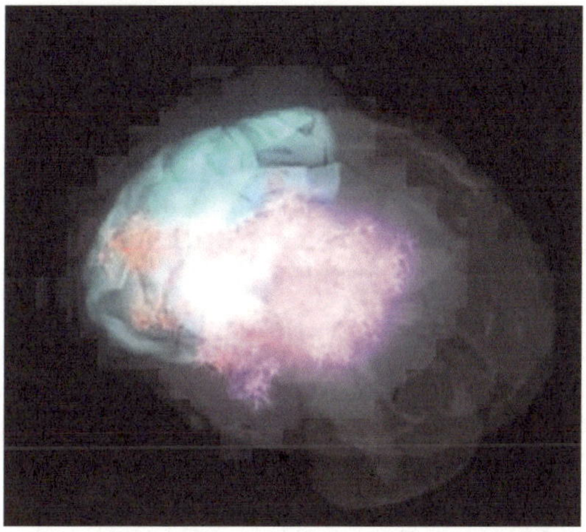

Figure 16.
fMRI area.

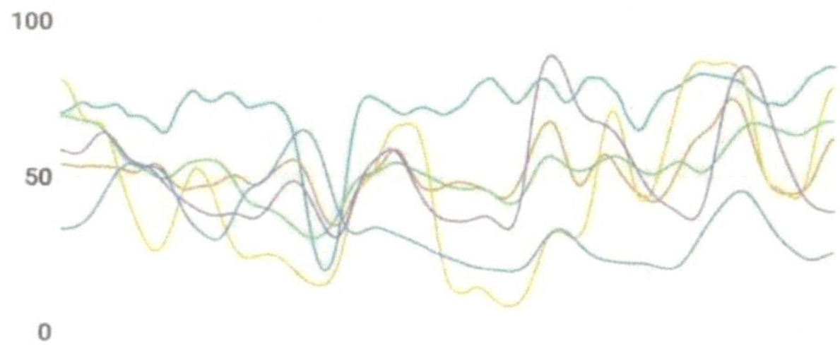

Figure 17.
EEG image.

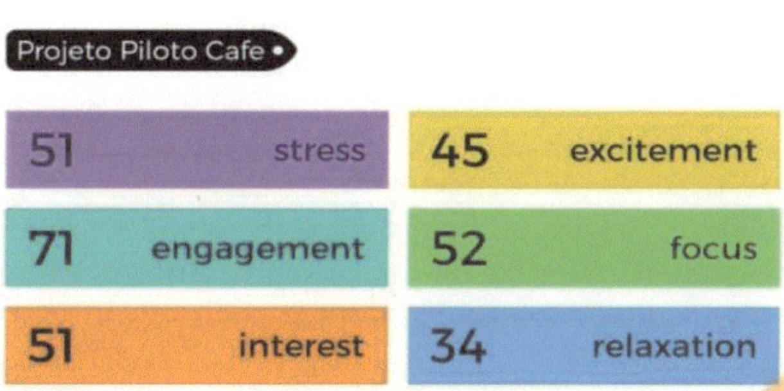

Figure 18.
EEG results.

prepared with a maximum of 500 mL of each sample and stored in thermos, remaining in place for a maximum of 90 minutes. The evaluation for aroma and flavor was made using 30 mL of each coffee sample served in transparent cups at an average temperature of 60 °C without the addition of sucrose. Between each sample, participants were served water at room temperature to wash the taste buds. Appearance was assessed using the conventional presentation (transparent cups and saucers) and transparency exists so that participants can see the color and texture of the drink. Visibility and light conditions of a common night in ambient lighting. For this experience, the environment used was an open area with room temperature and filming and protection equipment (mask, alcohol gel and gloves, in addition to receiving 1 person at a time in intervals of 3 to 4 hours between each participant to clean the place) The alone are served 2 times each, so that in each one the data of EEG + scale of SAATY and fMRI + saliva are collected.

Step 2: The equipment will be placed on the user and the user will taste the coffee and rate each sample using the SAATY scale, while the EEG data is generated. For measurement, the EPOC + equipment from the company Emotiv and the EEG software EmotivPro were used, which in addition to presenting the data for each frequency, generates the interpretation of these signals as stress, focus, motivation, engagement, relaxation, interest and excitement (**Figures 17** and **18**).

Step 3: The user will try it a second time, while the fRMI analysis is done. For measurement, the EPOC + equipment from Emotiv and the fMRI software BrainViz were used. During the test, 6 samples were taken from each user. Before drinking coffee, after the first sip and after the end of each cup. The saliva analyzes were stored to be sent to the laboratory.

Step 4: Movies and photos of people were stored in HD and cloud, as well as films from fMRI and EEG. Gel alcohol, masks and gloves were available at all times for participants with an isolated disposal site of at least 3 m.

Step 5: The linguistic terms of the SAATY scale (2.4 results) were developed and approved by food professionals such as restaurant owners, buffets, cheffs, housewives who cook for their families and even by a purchasing chef in a large chain. Supermarkets in the state of Rio de Janeiro in Brazil. The fMRI analyzes were made by the author with the cutout presented in brain response to food [20] and after adaptation, presented to specialists in neuroscience and consumer psychology at USP and UNIFESP for validation. EEG analyzes are taken from the software. The traditional EEG reading presents the frequencies in real time, however the software presents a reading (patented by the manufacturer) through the frequencies that generates responses of: interest, motivation, stress, engagement, focus and excitement.

Step 6: The data collected through fMRI, EEG and hormonal dosage in saliva are inserted in Matrix 2: Brain and hormones. The fMRI data is collected by analyzing the video of the brain at a speed of 30 frames per second, counting the moments of stimulus of the areas: Lateral Geniculate Body, Visual Cortex, Nucleus Accumbens, Olfactory Cortex, Frontal Cortex and standardized to be on a scale from 0 to 100, as is the answer presented by the EEG software. Inserted in the matrices, the data is processed through the algorithm created for this project and receives a final score (2.4.3 Verbal Analysis × Neural Analysis).

Step 7: Comparative analysis. After analyzing the samples of each participant, we have the verbal and neural score for each sample. These notes are compared in order to understand the preferred option of the participants in an individual and general way.

Step 8: Conclusion interview. After the analysis and conclusions, the participants of the experiment are invited to an interview where the variables involved and the preferences of the group are debated to verify points that influence the preference and which responses generate these variables. As we could see in the hamburger experiment (**Figures 4–6**), some answers can be contradictory, but this was not the case for this third experiment with the 3 coffee samples.

2.4 Results

The attributes selected for the option selection:

- Spit: Any reaction linked to the exclusion of food from the mouth;
- Disgust: I don't even want to try it;

- Bad;
- Indifferent: it is not bad, but it is also not good;
- Good: satisfied with the taste;
- Great - satisfied with the taste, he would eat again;
- Delicious - delicious.

2.4.1 Sensorial scale of verbal attributes

Before interpreting the data, a classification vector was constructed using the sets below (Dispensable: 1 to 3; Unsatisfactory: 3 to 4; Important: 4 to 5; Satisfactory: 5 to 7 and Crucial: 7 to 9). Where the attributes for grades 1 to 9 of the verbal scale (SAATY) are developed.

Set 1: Dispensable
if 1 < or = x, then u (x) = 1
if 2 < ou = x < ou = 3, then u (x) = − x + 3

Set 2: Unsatisfactory
if 2 < ou = x < ou = 3, then u (x) = x - 2
if 3 < ou = x < ou = 4, then u (x) = − x + 4

Set 3: Important
if 3 < ou = x < ou = 4, then u (x) = x - 3
if 4 < ou = x < ou = 5, then u (x) = − x + 5

Set 4: Satisfactory
if 4 < ou = x < ou = 5, then u (x) = x - 4
if 5 < ou = x < ou = 7, then u (x) = − x + 7

Set 5: Crucial
if 5 < ou = x < ou = 7, then u (x) = x - 7
if 7 < ou = x < ou = 9, then u (x) = 1

At the same time that the participant responds verbally, his brain is analyzed in fMRI, checking if the areas responsible for the feeding process, such as salivation, digestion and etc. are stimulated, and for how long the stimulus occurs in each area. To measure whether the participant's interest, involvement, excitement, focus, relaxation and stress, EEG equipment was used. Some of these states are linked to the production of hormones such as serotonin, endorphins and dopamines, among others, which are related to pleasure. These data generate an interference model to establish the criteria of the Fuzzy set of pertinence and can be crossed to generate a score for each experience with each coffee, which can be measured and compared.

In the table below, we can identify the sensation (column sensations) according to the linguistic terms (2.4 Results). The grades for each sample A, B and C are presented in the respective columns and the importance for the objective is applied by the researcher according to the sets above (which represent each grade).

The notes in **Table 1** are represented in the graph below (**Figure 13**). The graph in **Figure 14** represents the description of the Fuzzy Matrix, between the "linguistic terms" and the importance for the objective.

Sensations	Sample A	Sample B	Sample C	Importance for the objective
Split	1	1	1	Dispensable
Disgust	2	2	2	Dispensable
Bad	3	3	3	Dispensable
Doesn't matter	4	4	4	Unsatisfactory
Good	5	5	5	Important
Great	7	9	7	Satisfactory/crucial/satisfactory
Delicious	9	7	9	Crucial/satisfactory/crucial

Linguistic terms of importance for the objective: dispensable, unsatisfactory, important, satisfactory and crucial.

Table 1.
Verbal scale.

2.4.2 Sensorial scale of neural attributes

To interpret the neural matrix data, a new classification "Vector" is required, this time between 0 and 100, which was represented by the sets below (Dispensable: 10 to 30; Unsatisfactory: 30 to 40; Important: 40 to 50; Satisfactory: 50 to 70 and Crucial: 70 to 90 or greater). Where the attributes for the notes analyzed through the fMRI are developed (where we check 30 images per second to define how long each stimulus takes place, as well as making mathematical adjustments for quantification (each sample was tasted for 1 min, therefore, having the same time interval of each sample and the time in which each area was stimulated, we can interpret this data as a number between 0 and 100 that will fit perfectly in the neural matrix and in the Vector for interpretation of the algorithm (**Table 2**).

Set 1: Dispensable (g) = 10
c (g) = x

Set 2: Dispensable (g) = 20
10 < n < or = 20, n (g) = x + 10

Set 3: Dispensable (g) = 30
20 < r < or = 30, n (g) = x + 20

Set 4: Unsatisfactory (g) = 40
30 < i < ou = 40, n (g) = x + 30

Set 5: Important (g) = 50
40 < b < ou = 50, n (g) = x + 40

Set 6: Satisfactory (g) = 70
50 < o < ou = 70, n (g) = x + 50

Set 7: Crucial (g) = 90
70 < d < ou = 90, n (g) = x + 7

Stimuli	Sample A	Sample B	Sample C	Importance for the objective
Lateral geniculate body	76	86	64	Satisfactory/satisfactory/important
Visual cortex	52	80	58	Important/satisfactory/important
Nucleus accumbens	90	80	88	Crucial/satisfactory/satisfactory
Olfactory cortex	71	86	70	Satisfactory/satisfactory/satisfactory
Frontal cortex	76	80	76	Satisfactory/satisfactory/satisfactory

Stimuli	Sample A	Sample B	Sample C	Importance for the objective
Stress	51	51	43	Important/important/unsatisfactory
Excitement	45	36	22	Unsatisfactory/dispensable/dispensable
Engagement	71	73	77	Satisfactory/satisfactory/satisfactory
Focus	52	48	34	Important/unsatisfactory/dispensable
Interest	51	55	65	Important/important/important
Relaxation	34	31	30	Dispensable/dispensable/dispensable

Linguistic terms of importance for the objective: dispensable, unsatisfactory, important, satisfactory and crucial.

Table 2.
Neural stimuli.

The notes for stress, excitement, engagement, focus, interest and relaxation stimuli are interpreted and sent by the software. Below in **Figures 15** and **16**, we can see the fMRI images and the areas stimulated with different frequencies (**Figure 16**). In **Figure 17**, we can analyze the EEG data during the process of the experiment with one of the coffee samples, thus being able to understand peaks and falls of the stimuli, as well as identify what happened in the experiment that may have been responsible for the changes in the graph. **Figure 18** shows the results of the EEG software, it can be seen that the colors of each attribute are represented in the graph of **Figure 17**.

In **Figure 19**, we can see the graph of the Neural response and its interpretation is very similar to the graph in **Figure 13**, where the terms of importance for the objective are on the vertical axis and the stimulated areas, as well as the EEG attributes are positioned to give us a visual representation of the difference between samples for one participant.

2.4.3 Verbal analysis × neural analysis

The standardization of notes through vectors and the importance for the objective, aims to transform the numbers on a scale, where the largest number is the most pertinent, that is, the higher the number found in the analysis of the algorithm, the greater the relevance of that one. Sample regarding the participant's preference for the sample.

Sample A
Verbal = 1.4474 | Verbal Analysis Winner
Neural = 1.9886

Sample B
Verbal = 1.4447
Neural = 2.0515 | Neural Analysis Winner

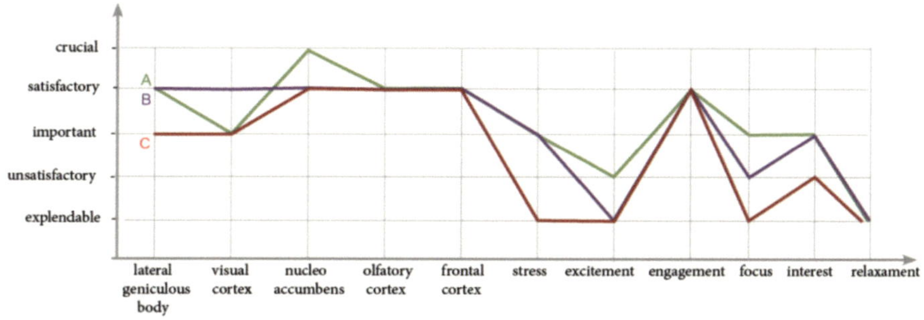

Figure 19.
Neural response graph, sample A, B and C.

Sample C
Verbal = 1.4447
Neural = 1.9834

2.4.4 Comparison between samples

Sample A | Verbal / Neural
(1.4474 + 1.9886) / 2 = 1.7178

Sample B | Verbal / Neural
(1.4447 + 2.0515) / 2 = 1.7481 | Winner of Verbal / Neural Analysis

Sample C | Verbal / Neural
(1.4447 + 1.9834) / 2 = 1.7140

3. Conclusion

A tie can be observed if verbal analysis is considered, but neural analysis showed preference for Sample B. For one more test, the average between the verbal/neural scores of each sample was made in order to be able to compare them to each other, giving Sample B the winner again. Considering previous experiences and those scheduled for the coming years, the multicriteria algorithm is developed according to the number of linguistic terms, areas to be analyzed and the number of participants. The linguistic terms vary according to the segment (food, perfumes, etc.) and experts must always be used so that the terminology reproduces the objectives of the experiment. In this case, the verbal scale was discussed and approved by professionals from different areas related to food and beverages, such as restaurant chefs, sommeliers, food and beverage entrepreneurs, as well as empirical specialists (people who cook for home or small businesses) and professional of a supermarket chain in Brazil. The neural matrix was built based on models and studies by CARNELL S. and Emotiv's EEG software.

3.1 Future reviews

This same data can be used to analyze peaks and falls individually and to understand what factors can be determined by these effects. Analysis of each element of a layout, product or service such as colors, images, typography, packaging, internal and external experience, etc. Thus, the same experience can be analyzed at satisfactory and unsatisfactory peak times, identifying them and using the information both to avoid unsatisfactory, and to praise and build their offers and experiences based on the observer's satisfaction, both for a product/service and for an environment/experience.

Author details

Alexandre Magno da Rocha Vianna
DNAhub Neurodesign, Rio de Janeiro, RJ, Brasil

*Address all correspondence to: alexandre@dna-neurodesign.com

References

[1] Simon H 1975, Prx00EA;mio Turing da ACM, juntamente com Allen Newell, pelas suas "contribuix00E7;x00F5;es bx00E1;sicas x00E0; Inteligx00EA;ncia Artificial, x00E0; Psicologia de Cognix00E7;x00E3;o Humana, e ao processamento de listas".

[2] Kahneman D, Tversky. Prospect Theory: An Analysis Of Decision Under Risk. Econometrica, Vol. 47, No. 2. 1979, pp. 263x2013;291. DOI: 10.2307/1914185

[3] Kotler P, Keller K, Administrax00E7;x00E3;o de Marketing. New Jersey: Pearson Prentice Hall. 2012, 12a ed.

[4] Lindstrom M. A Lx00F3;gica do consumo, 2016. Brasil: Harper Collins. ISBN-10: 8595082669

[5] Alexandre Magno da Rocha Vianna. Multicriteria Algorithms for Multisensory Design. Human Interaction, Emerging Technologies and Future Applications III. IHIET 2020. Advances in Intelligent Systems and Computing, vol 1253. Springer, Cham. DOI: 10.1007/978-3-030-55307-4_12

[6] Freud S 1915, The Unconscious, (Rio de Janeiro: Imago) vol 1, 98-136p; 1915.

[7] Maslow A H. Motivation and Personality. 1970, Nova Iorque: Editora Harper x0026; Row, Publishers. ISBN 0-06-041987-3

[8] Herzberg F, Mausner B, Snyderman B. The motivation to work. Transaction Publishers. New Brunswick (USA) and London (UK). Reprint Routledge, 2017. ISBN 1351504428, 9781351504423.

[9] Csikszentmihaly L's 1. Flow. The Psychology of Optimal Experience. HarperCollins Publishers, Inc 1996

[10] Norman D. Emotional Design: Why We Love Or Hate Everyday Things. 2005, New York: Basic Books.

[11] Krippendorff K 2004, Intrinsic Motivation and Human-Centered Design. Pennsylvania: University of Pennsylvania, Departamental Papers ASC. 10.1080/1463922031000086717

[12] Jordan P. Designing Pleasurable Products. Cleveland: CRC Press, 2000. ISBN-10: 0415298873

[13] Tanaka K 2003, Introdux00E7;x00E3;o aos conjuntos nebulosos. Tradux00E7;x00E3;o de Osvaldo R. Saavedra (GSP-DEE x2013;UFMA). DOI: 10.1002/3527601821.

[14] Ross T. Fuzzy Logic With Engineering Applications. 2010, Wiley: John Wiley x0026; Sons 3rd Edition. DOI: 10.1002/9781119994374

[15] Carter Rita. Mapping the mind. Edited by Cambridge University, 2000

[16] Robert S. Feldman, James A. Forrest x0026; Benjamin R. Happ (2002) Self-Presentation and Verbal Deception: Do Self-Presenters Lie More?, Basic and Applied Social Psychology, 24:2, 163x2013;170. DOI: 10.1207/S15324834BASP2402_8

[17] Silverthorn D U. Fisiologia Humana, uma abordagem Integrada 5x00B0; Ed. 2010.

[18] Dispenza J. Evolving your brain and the science of creating personal reality. 2007. ISBN-10: 0x2013;7573-97789

[19] Cosenza C A N. Toledo O M. Um caso de aplicax00E7;x00E3;o da Lx00F3;gica Fuzzy x2013; o Modelo Coppe-Cosenza de Hierarquia Fuzzy. XXIII Encontro Nac. de Eng. de Produx00E7;x00E3;o - Ouro Preto, MG, Brasil, ENEGEP 2003.

[20] CARNELL, Susan. SCHUR, Ellen. What Twin Studies Tell Us About Brain Responses to Food Cues. 2017.

Section 2

Clinical Application of Electroencephalography

Chapter 4

EEG Measurement as a Tool for Rehabilitation Assessment and Treatment

Hideki Nakano

Abstract

In recent years, neuroscience-based rehabilitation, also known as neurorehabilitation, has been attracting increasing attention worldwide. Electroencephalography (EEG) has been widely used in clinical practice as a tool for the evaluation and treatment of rehabilitation because of its noninvasive and simple measurement of human brain activity. EEG-electromyography coherence is a method to analyze the synchronization between the motor cortex and muscle activity during movement and to quantitatively assess how the motor cortex controls muscle activity. In addition, recent advances in analysis and measurement techniques have made it possible to estimate the source of EEG signals, thus serving as a method to evaluate rehabilitation. The brain-machine interface, which integrates medicine and engineering, has been widely applied in the treatment of rehabilitation and for improving the quality of life. This chapter provides an overview of EEG, and its uses as a tool for rehabilitation assessment and treatment.

Keywords: EEG, non-invasive brain function measurement, rehabilitation, EEG–EMG coherence, EEG mapping, EEG source imaging, brain-machine interface

1. Introduction

Ever since the first human electroencephalography (EEG) [1] and electromyography (EMG) [2] recordings were performed in the 1920s, the theoretical aspects, test techniques, and clinical applications of each have rapidly advanced [3]. Methods for imaging brain function have appeared one after another over the past century beginning with evoked potentials [4] in the 1940s, event-related potential [5, 6] and magnetoencephalography (MEG) [7] in the 1960s, positron emission tomography (PET) [8, 9] in the 1970s, and functional magnetic resonance imaging (fMRI) [10, 11] in the 1990s. Currently, the noninvasive methods available for measuring brain function are broadly divided into two categories: electrophysiological examinations and imaging techniques based on hemodynamic principles. The former includes EEG, MEG, and transcranial magnetic stimulation (TMS), while the latter includes fMRI, PET, single photon emission computed tomography (SPECT), and near-infrared spectroscopy (NIRS) [12].

EEG is widely used in rehabilitation as it is well suited to the field's demands for measurement, which includes simple, safe, and portable equipment. In the past, EEG has primarily been an analysis method used to capture brain activity accompanying a given phenomenon or during a given task as an electric field and subsequently estimates the source of that activity based on the distribution on the scalp. In contrast, recent advancements have led to the development of a method capable of capturing fluctuations in the power of rhythms in a certain frequency band. When this power decreases accompanying a given phenomenon or task, it is called event-related desynchronization (ERD). Conversely, when this power increases accompanying a given phenomenon or task, it is called event-related synchronization (ERS) [13–15]. Thus, electric field analysis is an analysis of the temporal domain, while the second method is an analysis of the frequency domain. In frequency analysis, ERD is thought to reflect a state of increased cortical activity in the region, while ERS is thought to reflect a state of decreased activity or return to a low level. This chapter will outline the clinical applications for treatment and evaluation of rehabilitation using these features of EEG focusing specifically on EEG–EMG coherence, scalp mapping, and brain-machine interface.

2. EEG-EMG coherence

Like brain waves, it has long been known that myoelectric activity—the final output of the motor system—is rhythmic. Since a correlation between EEG and EMG rhythms was first reported, the concept of EEG–EMG coherence has become a field of study attracting much attention [16–18]. As EMG measures the collective firing of a motor unit, if rectified such that the positivity or negativity of individual spikes is irrelevant, EMG signals are thought to correspond to action potentials of spinal motor neurons [19]. At the same time, EEG activity reflects the collective activity of neurons, particularly their postsynaptic potential. Therefore, EEG–EMG coherence is considered capable of measuring the control of spinal motor neurons by the cerebral cortex.

In healthy individuals, EEG–EMG coherence shows a distribution following the somatotopy of the primary sensorimotor cortex contralateral to the muscle for which myoelectric activity was recorded. Research using MEG has found that the source of coherent rhythmic activity can be found in the primary motor cortex [20, 21]. Further, peak coherence has been reported to roughly correspond to hot spots during TMS [17]. Significant coherence is primarily seen in the β frequency band (13–30 Hz) but has also been observed in the lower frequency α band and the γ band near 40 Hz. Thus, coherence in these various frequency bands may be derived from different mechanisms [22].

Research measuring the time lag between EEG and EMG has found that EEG invariably precedes EMG for the β band, yet there is almost no lag for the α band [18]. This suggests that the mechanisms of coherence in the α and β bands differ. One theory to explain this is that a muscle's peripheral centrifugal sensory input is involved in α band coherence. However, a previous study found that coherence in this band was not affected when peripheral sensory input was modified using vibration stimulation [18]. Thus, it appears that the reason there is no time lag between cortical activity and myoelectric activity is that subcortical rhythmic activity contributes to both brain wave rhythms and myoelectric activity. Further, studies have found that intensifying muscle contraction changes the coherence peak frequency from the β band to the γ band [23, 24]. This γ band coherence is thought to contribute to the control of myoelectric activity (piper rhythm) at approximately 40 Hz, as is seen during strong muscle contraction. Interestingly, the coherence peak does not transition smoothly from the β band to the γ band as myoelectric activity changes from weak contraction to strong contraction; rather, it shifts in a step-like manner. This suggests that the mechanism involved in coherence in the γ band differs from that of the β band. However, there is no difference between the two frequency bands when measuring the time lag between brain activity and myoelectric activity; brain activity precedes myoelectric activity for both. Accordingly, coherence in both of these frequency bands is thought to be involved in centrifugal output from the cerebral cortex to spinal motor neurons. This type of coherence is localized to the primary sensorimotor cortex contralateral to the muscle. However, subdural recordings of patients with intractable epilepsy requiring surgical intervention have shown EEG–EMG coherence in other brain areas, such as the premotor cortex and supplementary motor cortex [25]. Anatomically, its well known that there are direct projections from the premotor cortex and supplementary motor cortex to spinal motor neurons [26, 27], suggesting that these brain areas are involved in the control of myoelectric activity.

Due to its ability to non-invasively measure frequency-specific coupling of the cerebral cortex (specifically the primary motor cortex) and spinal motor neurons, clinical applications of EEG–EMG coherence are ongoing and include illuminating the pathophysiology and evaluation of diseases featuring motor impairment or involuntary movement. A relatively slow resting tremor of 3–6 Hz is one of the core symptoms of Parkinson's disease. While the rhythm of these tremors is thought to originate in the basal ganglia-thalamo-cortical loop, the mechanism of onset remains unknown. One study exploring the EEG–EMG coherence of these resting tremors found that primary sensorimotor cortex activity corresponds to the tremors [28]. As Parkinson's disease patients exhibit EEG–EMG coherence at their tremors' peak frequency or double harmonic frequency, stronger coherence is observed between 5 and 12 Hz, a range that displays low coherence in healthy individuals. At the same time, such patients show reduced coherence in other frequency bands (15–60 Hz) [29]. This abnormal coherence pattern has been found to approach that of healthy individuals (strong coherence for 15–60 Hz) with the use of deep brain stimulation or pharmacotherapy using drugs such as levodopa [30, 31]. Thus, the dopaminergic system may influence the occurrence of this coherence. Studies also report EEG–EMG coherence features resembling those of resting tremors in relation to freezing of gait, a typical gait disorder seen in patients with Parkinson's disease [32, 33]. Accordingly, EEG–EMG coherence is considered widely applicable as a tool for evaluating the effects of rehabilitation interventions and elucidating the pathology of movement disorders in patients with Parkinson's disease.

Reduced EEG–EMG coherence has been reported not only in patients with Parkinson's disease, but also in stroke patients and older adults. One study examining the EEG–EMG coherence of the hemiplegic and non-hemiplegic sides of subcortical infarction patients found that although the EEG and EMG power showed similar patterns for both the hemiplegic and non-hemiplegic sides, the coherence was significantly lower on the hemiplegic side [34]. This reduced EEG–EMG coherence on the hemiplegic side has been shown to improve as the patient's motor function recovers [35], suggesting that this may be a useful biomarker reflecting motor function recovery in stroke patients. Meanwhile, EEG–EMG coherence in older adults is significantly lower than that in younger individuals and has been shown to have a significant correlation with muscle strength [36]. This suggests that lower EEG–EMG coherence in older adults may be one factor in the decline in strength, motor skills, and coordination that accompanies aging.

3. EEG mapping and source imaging

Interpreting an EEG visually requires experience, but a two-dimensional representation of brain electrical activity (topography) is a way to display brain waves more objectively as a planar map of electrical activity on the scalp's surface. Techniques are also being developed to estimate areas of activity within the brain from multichannel EEG data obtained from the scalp, thereby increasing the precision of brain function analysis using EEG.

EEG scalp mapping analyses include spatial analysis (two-dimensional and three-dimensional), coherence, and complexity (Ω). As brain waves consist of multiple frequencies with different physiological significances, it is vital to perform frequency analysis based on a fast Fourier transform (FFT) to consider each frequency independently. It is also integral to select the appropriate analysis and interpretation with consideration to the items to be evaluated and features of each disease using these analysis techniques [37].

A previous study reported the spatial distribution of EEG topography independent of electrode placement by epoch and found that the standard topographies of various intervals were separated by instantaneous transitions [38]. In other words, it was unusual for one shape to slowly change into the next. Different topographies are thought to reflect different regions of neural activity and represent different stages of information processing. In light of this, dividing brain waves according to the temporal similarity of their spatial distribution on the scalp is considered a potentially useful method for studying information processing within the brain as it changes moment to moment. EEG microstate modeling and analysis was developed as a method of microstate segmentation using cluster analysis to determine the optimal topography and number of segments from a sequence of brain electrical activity corresponding to the characteristics of a mental activity [39]. This method is used to efficiently extract data based on the temporal and spatial structure of background EEG activity and explore the pathophysiology of brain function in a number of diseases [40, 41].

Importantly, a three-dimensional approach is necessary when considering actual brain pathology. Estimating the source of brain waves has recently been gaining attention as one approach to three-dimensional EEG analysis. This approach can be broadly divided into equivalent dipole estimation methods [42–44] and low-resolution brain electromagnetic tomography (LORETA) [45–47], a standard method of current density distribution estimation. While there are advantages and disadvantages to each, one challenge faced by the former, for which it is essential to stipulate the number of sources of activity in advance, is the difficulty of selecting which combination of dipoles is valid because different combinations of dipoles result in similar scalp distributions (inverse problem). The latter depicts the spread of neural activity within the brain in three-dimensional tomography using EEG data collected from the scalp based on the hypothesis that adjacent groups of neurons have roughly the same activity. Excluding special cases such as epileptic seizures, actual brain activity is not limited to one specific area, making this method useful in understanding complex brain activity such as higher brain function. More specifically, LORETA excels in primary processing, analyzing raw data to display an image, and secondary processing, carrying out statistical analyses to extract maps and find differences in current density distributions, and is therefore a form of EEG mapping used in diverse branches of neuroscience. As discussed above, LORETA estimates a three-dimensional distribution of brain tissue activity from EEG data measured on the scalp based on the hypothesis that adjacent neuron groups carry out similar activity. In other words, assuming a number of cubic lattices within the cerebral parenchyma, this method generates a three-dimensional blurred image of the current source by selecting the smoothest option from among combinations of three-dimensional current density distributions based on the Laplacian operation. Unlike other programs, the initial location value or number of dipoles is not set in advance. The operation is relatively simple, and while the resolution is low, the result

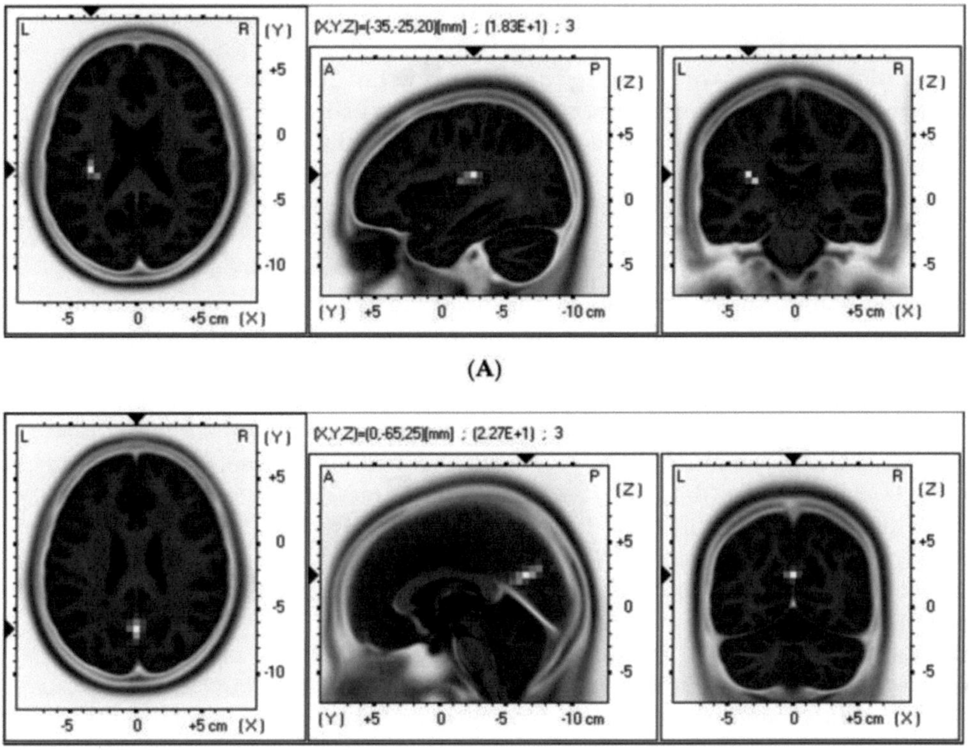

Figure 1.
Statistical non-parametric maps of LORETA of the alpha band comparing pre-rest and post-rest of hand massage (A) and foot massage (B) [48]. White blobs indicate increased activity at post-rest for each massage.

is not a primitive spherical model, but instead a tomographic image superimposed onto Talairach atlas, which can be shown in color and three dimensions (**Figure 1**) [48–50]. LORETA is being improved, and it has recently become possible to evaluate functional lagged connectivity and the directionality of that connectivity (isolated effective coherence; iCoh) between different areas of the brain.

As demonstrated above, delving deeper into background EEG activity by first exploring the time domain using methods such as microstate segmentation, then investigating the frequency domain using FFT, and the spatial domain both two-dimensionally (topography) and three-dimensionally (equivalent dipole estimation, FFT-dipole-approximation, LORETA) has a wide range of clinical applications, including elucidating pathological mechanisms and evaluating rehabilitation.

4. Brain-machine interface (BMI)

BMI techniques are methods of connecting the exchange of information between the external world and the brain using artificial electric circuits to restore and supplement its function. In the field of rehabilitation, output-type BMI applications, which read motor intention from brain activity and use this information to operate various devices and computers, are commonly used. Output-type BMI, which interprets motor intention from brain activity to operate external equipment, is classified into invasive and noninvasive types based on the method by which brain activity is measured. The former uses intracranial or epidural electrodes; the latter uses scalp EEG or functional brain imaging techniques. In addition to the conventional methods of restoring function using BMI, such as directly operating a robot arm or environmental control apparatus using brain activity, research geared toward therapeutic BMI applications, which utilize BMI for rehabilitation or reconstruction of functional neural networks, is also underway.

Neurofeedback is a method of learning to voluntarily control one's own brain activity through the presentation of said activity as real-time sensory information (visual, auditory, etc.) (**Figure 2**) [51]. Neurofeedback requires technology that measures brain activity and analyzes the measured data in real time. The technologies involved in brain signal processing and interpretation are shared with those of BMI and, in a broad sense, neurofeedback can be considered a therapeutic form of BMI. In fact, EEG-based neurofeedback is widely used as a tool for improving motor, cognitive, and psychological functions not only in individuals with diseases, but also in healthy individuals ranging from childhood to old age. The delta (<4 Hz), theta (4–8 Hz), alpha (8–13 Hz), beta (14–30 Hz), and gamma (>40 Hz) frequency bands are most commonly used in evaluation and training [52]. As the functional characteristics of each frequency band differ, it is essential to select the appropriate frequency band for neurofeedback depending on the pathology of the case or the type of function one wishes to improve (**Table 1**) [53].

Neurofeedback is also gaining popularity as a technique for neuromodulation, that is, the regulation of local brain activity. Neurofeedback is considered very safe compared to methods such as repetitive transcranial magnetic stimulation (rTMS) or transcranial direct current stimulation (tDCS), as it does not use external stimulation and therefore avoids the risk of side effects such as seizure or burns that occur with rTMS and tDCS. At the same time, output-type BMI has been gaining interest in recent years as a tool for supporting the daily activities of persons who have difficulty with independent living or spontaneous expression due to disease, disability, or aging. Specifically, it will soon become possible to operate a variety of assistive devices, including wheel chairs, exoskeletons, drones, and communication robots using the operator's EEG signals (**Figure 3**) [54]. Researchers are also

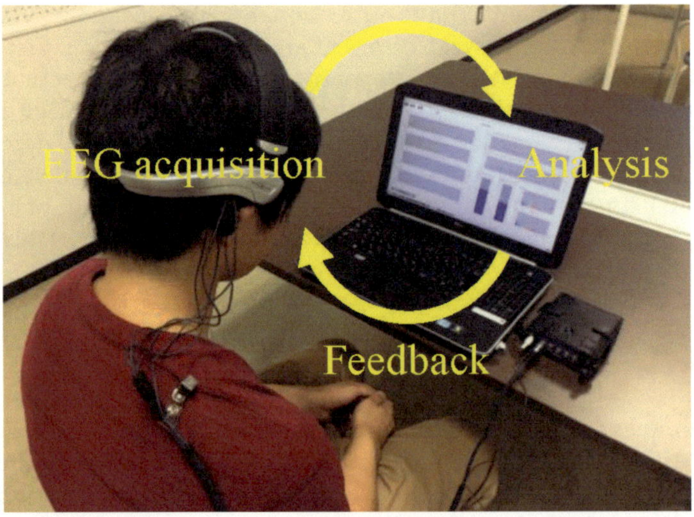

Figure 2.
Motor imagery training using neurofeedback, a therapeutic BMI [51]. EEG activity feedback during motor imagery (μ band EEG activity in the sensorimotor area) is given using sensory modalities such as vision or hearing and the participant is trained to control their own EEG activity.

Protocol	Purpose
↓ theta	Cognitive training after stroke;
	Cognitive training of healthy adults with a risk for neurodegenerative disorder.
↑ theta	Aiming to increase capabilities of executive functions on healthy students;
	Memory consolidation training.
↑ theta, ↓ alpha	Relaxation training;
	Training to improve creative performance (playing music, dancing), effects on mood.
↓ alpha	Attentional training;
	Frontal alpha-asymmetry self-regulation training to influence mood;
	Training for increased motor performance.
↑ alpha;	Training to reduce anxiety;
	Training to improve cognitive performance;
	Relaxation training for stress reduction.
↑ high alpha	Training to improve cognitive performance.
↑ SMR (12-15 Hz)	Training to decrease epileptic seizures;
	Training to improve declarative learning and sleeping pattern;
	Training to improve cognition and memory in stroke patients;
	Training to enhance golf putting.
↑ SMR, ↓ theta	Training to optimize microsurgical skills;
	Training to minimize ADHD symptoms on a healthy population.
↑ SMR, ↓ theta, ↓ high beta	Training to improve cognitive performance;
	Training to improve Asperger's syndrome and autistic spectrum disorder symptoms.
↑ low beta	Training to improve cognitive performance;
	Training to modulate sleep spindle activity and overnight memory consolidation.
↑ beta, ↓ theta	Typical training for improvement of ADHD symptoms.
↑ beta, ↓ theta, ↓ low alpha	Training of attention.
↑ gamma	Training of cognitive control;
	Training of memory and intelligence.

Table 1.
An overview of already used protocols of frequency EEG-neurofeedback training with the references to exemplary studies and their main therapeutic purpose [53].

developing and exploring the effectiveness of smart homes that incorporate these BMI technologies [55]. Smart homes are equipped with technology that interprets the user's motion intention or emotional state using methods such as EEG, which can easily measure brain activity with no special training or burden on the user. Specifically, smart homes assist with daily life by measuring the brain activity that occurs when the user naturally moves their body accompanying a motion intention, for example, to operate the television or air conditioner, recognizing what kind of motion intention is occurring, and manipulating the environment in accordance with the user's intention. They may also detect when the user is feeling discomfort and modify the environment accordingly using technology that captures an

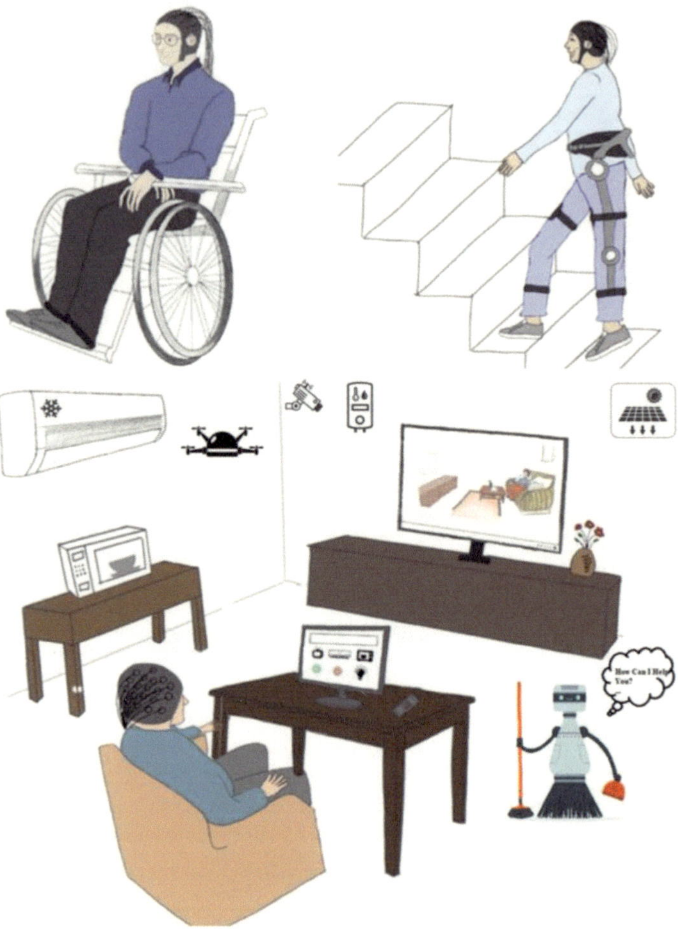

Figure 3.
EEG-based BMI for assisting with daily activities and improving quality of life [54].

emotional state (discomfort) by measuring and analyzing the user's brain activity. This information can further be communicated to family members or caregivers to allow them to provide assistance based on the user's emotional state. In addition to the above, it is also possible to assist a user's own actions in a standard living environment using BMI actuation technology that moves an exoskeleton-type robot actuator linked to brain activity [54, 55]. It is hoped that such BMI technologies will increase communication in a variety of settings and create an environment where people can continue to live independent fulfilled lives.

5. Conclusion

The development of brain function imaging techniques has led to a new understanding of previously unexplained brain functions as well as the creation of clinical applications, scientific techniques, and assistive devices based on these findings. Elucidation of brain function fully utilizing the advantages of EEG is expected to continue as high definition EEG and accompanying analysis methods become more advanced. EEG is also advantageous in that it is relatively easy to record simultaneously with other methods of brain function measurement. Therefore, it is imperative that we do not simply interpret pathology and brain function using EEG results alone but gain a comprehensive picture of the brain's physiological function and dysfunction through the simultaneous use of multiple methods of brain function measurement and by capturing various clinical parameters in a multidimensional manner.

Acknowledgements

This research was funded by the Japanese Physical Therapy Association.

Conflict of interest

The authors declare no conflict of interest.

Appendices

EEG	electroencephalography
EMG	electromyography
MEG	magnetoencephalography
PET	positron emission tomography
fMRI	functional magnetic resonance imaging
TMS	transcranial magnetic stimulation
SPECT	single photon emission computed tomography
NIRS	near-infrared spectroscopy
ERD	event-related desynchronization
ERS	event-related synchronization
FFT	fast Fourier transform
LORETA	low-resolution brain electromagnetic tomography
BMI	brain-machine interface
rTMS	repetitive transcranial magnetic stimulation
tDCS	transcranial direct current stimulation
SMR	sensorimotor rhythm
ADHD	attention-deficit hyperactivity disorder

Author details

Hideki Nakano
Neurorehabilitation laboratory, Graduate School of Health Sciences,
Kyoto Tachibana University, Kyoto, Japan

*Address all correspondence to: nakano-h@tachibana-u.ac.jp

References

[1] Berger H. Über das Elektrenkephalogramm des Menschen. Archiv f. Psychiatrie. 1929;**87**:527-570. DOI: 10.1007/BF01797193

[2] Adrian ED, Bronk DW. The discharge of impulses in motor nerve fibres: Part II. The frequency of discharge in reflex and voluntary contractions. The Journal of Physiology. 1929;**67**(2):i3-i151

[3] Shibasaki H. Current status and future of non-invasive studies of human brain functions. Japanese Journal of Clinical Neurophysiology. 2008;**36**(3):114-121

[4] Dawson GD. Investigations on a patient subject to myoclonic seizures after sensory stimulation. Journal of Neurology, Neurosurgery, and Psychiatry. 1947;**10**(4):141-162. DOI: 10.1136/jnnp.10.4.141

[5] Walter WG, Cooper R, Aldridge VJ, Mccallum WC, Winter AL. Contingent negative variation: An electric sign of sensorimotor association and expectancy in the human brain. Nature. 1964;**203**:380-384. DOI: 10.1038/203380a0

[6] Sutton S, Braren M, Zubin J, John ER. Evoked-potential correlates of stimulus uncertainty. Science. 1965;**150**(3700):1187-1188. DOI: 10.1126/science.150.3700.1187

[7] Cohen D. Magnetoencephalography: Evidence of magnetic fields produced by alpha-rhythm currents. Science. 1968;**161**(3843):784-786. DOI: 10.1126/science.161.3843.784

[8] Orgogozo JM, Larsen B. Activation of the supplementary motor area during voluntary movement in man suggests it works as a supramotor area. Science. 1979;**206**(4420):847-850. DOI: 10.1126/science.493986

[9] Roland PE, Larsen B, Lassen NA, Skinhøj E. Supplementary motor area and other cortical areas in organization of voluntary movements in man. Journal of Neurophysiology. 1980;**43**(1):118-136. DOI: 10.1152/jn.1980.43.1.118

[10] Ogawa S, Lee TM. Magnetic resonance imaging of blood vessels at high fields: in vivo and in vitro measurements and image simulation. Magnetic Resonance in Medicine. 1990;**16**(1):9-18. DOI: 10.1002/mrm.1910160103

[11] Ogawa S, Lee TM, Kay AR, Tank DW. Brain magnetic resonance imaging with contrast dependent on blood oxygenation. Proceedings of the National Academy of Sciences of the United States of America. 1990;**87**(24):9868-9872. DOI: 10.1073/pnas.87.24.9868

[12] Shibasaki H. Human brain mapping: Hemodynamic response and electrophysiology. Clinical Neurophysiology. 2008;**119**(4): 731-743. DOI: 10.1016/j.clinph.2007.10.026

[13] Pfurtscheller G, Aranibar A. Event-related cortical desynchronization detected by power measurements of scalp EEG. Electroencephalography and Clinical Neurophysiology. 1977;**42**(6):817-826. DOI: 10.1016/0013-4694(77)90235-8

[14] Pfurtscheller G. The cortical activation model (CAM). Progress in Brain Research. 2006;**159**:19-27. DOI: 10.1016/S0079-6123(06)59002-8

[15] Pfurtscheller G, Lopes da Silva FH. Event-related EEG/MEG synchronization and desynchronization: Basic principles. Clinical Neurophysiology. 1999;**110**(11):1842-1857. DOI: 10.1016/s1388-2457(99)00141-8

[16] Mima T, Shibasaki H. EEG-EMG coherence: Its application and limitation. Clinical Electroencephalography. 2003;**45**(1):19-25

[17] Mima T, Hallett M. Corticomuscular coherence: A review. Journal of Clinical Neurophysiology. 1999;**16**(6):501-511. DOI: 10.1097/00004691-199911000-00002

[18] Mima T, Steger J, Schulman AE, Gerloff C, Hallett M. Electroencephalographic measurement of motor cortex control of muscle activity in humans. Clinical Neurophysiology. 2000;**111**(2):326-337. DOI: 10.1016/s1388-2457(99)00229-1

[19] Halliday DM, Rosenberg JR, Amjad AM, Breeze P, Conway BA. Farmer SF. A framework for the analysis of mixed time series/point process data-theory and application to the study of physiological tremor, single motor unit discharges and electromyograms. Progress in Biophysics and Molecular Biology. 1995;**64**(2-3):237-278. DOI: 10.1016/s0079-6107(96)00009-0.

[20] Salenius S, Portin K, Kajola M, Salmelin R, Hari R. Cortical control of human motoneuron firing during isometric contraction. Journal of Neurophysiology. 1997;**77**(6):3401-3405. DOI: 10.1152/jn.1997.77.6.3401

[21] Gross J, Kujala J, Hamalainen M, Timmermann L, Schnitzler A, Salmelin R. Dynamic imaging of coherent sources: Studying neural interactions in the human brain. Proceedings of the National Academy of Sciences of the United States of America. 2001;**98**(2):694-699. DOI: 10.1073/pnas.98.2.694

[22] Mima T, Hallett M, Shibasaki H. Coherence, cortico-muscular. In: Hallett M, editor. Handbook of Clinical Neurophysiology. Elsevier; 2003. pp. 87-94. DOI: 10.1016/S1567-4231(09)70155-8

[23] Brown P, Salenius S, Rothwell JC, Hari R. Cortical correlate of the piper rhythm in humans. Journal of Neurophysiology. 1998;**80**(6):2911-2917. DOI: 10.1152/jn.1998.80.6.2911

[24] Mima T, Simpkins N, Oluwatimilehin T, Hallett M. Force level modulates human cortical oscillatory activities. Neuroscience Letters. 1999;**275**(2):77-80. DOI: 10.1016/s0304-3940(99)00734-x

[25] Ohara S, Nagamine T, Ikeda A, Kunieda T, Matsumoto R, Taki W, et al. Electrocorticogram-electromyogram coherence during isometric contraction of hand muscle in human. Clinical Neurophysiology. 2000;**111**(11):2014-2024. DOI: 10.1016/s1388-2457(00)00448-x

[26] Dum RP, Strick PL. Spinal cord terminations of the medial wall motor areas in macaque monkeys. The Journal of Neuroscience. 1996;**16**(20):6513-6525. DOI: 10.1523/JNEUROSCI.16-20-06513.1996

[27] He SQ, Dum RP, Strick PL. Topographic organization of corticospinal projections from the frontal lobe: Motor areas on the medial surface of the hemisphere. The Journal of Neuroscience. 1995;**15**(5 Pt 1):3284-3306. DOI: 10.1523/JNEUROSCI.15-05-03284.1995

[28] Hellwig B, Häussler S, Lauk M, Guschlbauer B, Köster B, Kristeva-Feige R, et al. Tremor-correlated cortical activity detected by electroencephalography. Clinical Neurophysiology. 2000;**111**(5):806-809. DOI: 10.1016/s1388-2457(00)00248-0

[29] Salenius S, Avikainen S, Kaakkola S, Hari R, Brown P. Defective cortical drive to muscle in Parkinson's disease and its improvement with levodopa. Brain 2002;125(Pt 3):491-500. DOI: 10.1093/brain/awf042.

[30] Brown P, Marsden J, Defebvre L, Cassim F, Mazzone P, Oliviero A, et al. Intermuscular coherence in Parkinson's disease: Relationship to bradykinesia. Neuroreport. 2001;**12**(11):2577-2581. DOI: 10.1097/00001756-200108080-00057

[31] Marsden J, Limousin-Dowsey P, Fraix V, Pollak P, Odin P, Brown P. Intermuscular coherence in Parkinson's disease: Effects of subthalamic nucleus stimulation. Neuroreport. 2001;**12**(6):1113-1117. DOI: 10.1097/00001756-200105080-00013

[32] Günther M, Bartsch RP, Miron-Shahar Y, Hassin-Baer S, Inzelberg R, Kurths J, et al. Coupling between leg muscle activation and EEG during Normal walking, intentional stops, and freezing of gait in Parkinson's disease. Frontiers in Physiology. 2019;**10**:870. DOI: 10.3389/fphys.2019.00870

[33] Roeder L, Boonstra TW, Kerr GK. Corticomuscular control of walking in older people and people with Parkinson's disease. Scientific Reports. 2020;**10**(1):2980. DOI: 10.1038/s41598-020-59810-w

[34] Mima T, Toma K, Koshy B, Hallett M. Coherence between cortical and muscular activities after subcortical stroke. Stroke. 2001;**32**(11):2597-2601. DOI: 10.1161/hs1101.098764

[35] Krauth R, Schwertner J, Vogt S, Lindquist S, Sailer M, Sickert A, et al. del R. Millán J, Hinrichs H, Heinze HJ, Sweeney-Reed CM. Cortico-muscular coherence is reduced acutely post-stroke and increases bilaterally during motor recovery: A pilot study. Frontiers in Neurology. 2019;**10**:126. DOI: 10.3389/fneur.2019.00126

[36] Bayram MB, Siemionow V, Yue GH. Weakening of Corticomuscular signal coupling during voluntary motor action in aging. The Journals of Gerontology. Series A, Biological Sciences and Medical Sciences. 2015;**70**(8):1037-1043. DOI: 10.1093/gerona/glv014

[37] Tanaka H. Brain mapping of electroencephalogram: Clinical application. Japanese Journal of Clinical Neurophysiology. 2008;**36**(6):658-666

[38] Lehmann D. EEG assessment of brain activity: Spatial aspects, segmentation and imaging. International Journal of Psychophysiology. 1984;**1**(3):267-276. DOI: 10.1016/0167-8760(84)90046-1

[39] Koenig T, Lehmann D, Merlo MC, Kochi K, Hell D, Koukkou M. A deviant EEG brain microstate in acute, neuroleptic-naive schizophrenics at rest. European Archives of Psychiatry and Clinical Neuroscience. 1999;**249**(4):205-211. DOI: 10.1007/s004060050088

[40] Khanna A, Pascual-Leone A, Michel CM, Farzan F. Microstates in resting-state EEG: Current status and future directions. Neuroscience and Biobehavioral Reviews. 2015 Feb;**49**:105-113. DOI: 10.1016/j.neubiorev.2014.12.010

[41] Michel CM, Koenig T. EEG microstates as a tool for studying the temporal dynamics of whole-brain neuronal networks: A review. Neuroimage. 2018;180(Pt B):577-593. DOI: 10.1016/j.neuroimage.2017.11.062.

[42] He B, Musha T. Equivalent dipole estimation of spontaneous EEG alpha activity: Two-moving dipole approach. Medical & Biological Engineering & Computing. 1992;**30**(3):324-332. DOI: 10.1007/BF02446971

[43] Bai X, He B. On the estimation of the number of dipole sources in EEG source localization. Clinical Neurophysiology. 2005;**116**(9):2037-2043. DOI: 10.1016/j.clinph.2005.06.001

[44] Musha T, Okamoto Y. Forward and inverse problems of EEG dipole localization. Critical Reviews in Biomedical Engineering. 1999;**27**(3-5):189-239

[45] Pascual-Marqui RD, Michel CM, Lehmann D. Low resolution electromagnetic tomography: A new method for localizing electrical activity in the brain. International Journal of Psychophysiology. 1994;**18**(1):49-65. DOI: 10.1016/0167-8760(84)90014-x

[46] Pascual-Marqui RD. Standardized low-resolution brain electromagnetic tomography (sLORETA): technical details. Methods and Findings in Experimental and Clinical Pharmacology. 2002;24 Suppl D:5-12.

[47] Pascual-Marqui RD, Lehmann D, Koukkou M, Kochi K, Anderer P, Saletu B, et al. Assessing interactions in the brain with exact low-resolution electromagnetic tomography. Philosophical Transactions of the Royal Society A. 2011;**369**(1952):3768-3784. DOI: 10.1098/rsta.2011.0081

[48] Nakano H, Kodama T, Ueda T, Mori I, Tani T, Murata S. Effect of hand and foot massage therapy on psychological factors and EEG activity in elderly people requiring long-term care: A randomized cross-over study. Brain Sciences. 2019;**9**(3):54. DOI: 10.3390/brainsci9030054

[49] Nakano H, Kodama T, Sakamoto M, Ueda T, Tani T, Mori I, et al. Effect of hand massage on occupational leg swelling and resting-state electroencephalographic activity: A randomized cross-over study. International Journal of Clinical Research & Trials. 2018;**3**:125. DOI: 10.15344/2456-8007/2018/125

[50] Nakano H, Osumi M, Ueta K, Kodama T, Morioka S. Changes in electroencephalographic activity during observation, preparation, and execution of a motor learning task. The International Journal of Neuroscience. 2013;**123**(12):866-875. DOI: 10.3109/00207454.2013.813509

[51] Nakano H, Kodama T, Murata S, Nakamoto T, Fujihara T, Ito Y. Effect of auditory neurofeedback training on upper extremity function and motor imagery ability in a stroke patient: A single case study. International Journal of Clinical Research & Trials. 2018;**3**:126. DOI: 10.15344/2456-8007/2018/126

[52] Noachtar S, Binnie C, Ebersole J, Mauguière F, Sakamoto A. Westmoreland B. a glossary of terms most commonly used by clinical electroencephalographers and proposal for the report form for the EEG findings. The international federation of clinical neurophysiology. Electroencephalography and Clinical Neurophysiology. Supplement. 1999;**52**:21-41

[53] Omejc N, Rojc B, Battaglini PP, Marusic U. Review of the therapeutic neurofeedback method using electroencephalography: EEG neurofeedback. Bosnian Journal of Basic Medical Sciences. 2019;**19**(3):213-220. DOI: 10.17305/bjbms.2018.3785.

[54] Belkacem AN, Jamil N, Palmer JA, Ouhbi S, Chen C. Brain computer interfaces for improving the quality of life of older adults and elderly patients. Frontiers in Neuroscience. 2020;**14**:692. DOI: 10.3389/fnins.2020.00692

[55] Jamwal R, Callaway L, Winkler D, Farnworth L, Tate R. Evaluating the use of smart home technology by people with brain impairment: Protocol for a single-case experimental design. JMIR Research Protocols. 2018;7(11):e10451. DOI: 10.2196/10451

Chapter 5

Fronto-Temporal Analysis of EEG Signals of Patients with Depression: Characterisation, Nonlinear Dynamics and Surrogate Analysis

Subha D. Puthankattil

Abstract

The recent advances in signal processing techniques have enabled the analysis of biosignals from brain so as to enhance the predictive capability of mental states. Biosignal analysis has been successfully used to characterise EEG signals of unipolar depression patients. Methods of characterisation of EEG signals and the use of nonlinear parameters are the major highlights of this chapter. Bipolar frontopolar-temporal EEG recordings obtained under eyes open and eyes closed conditions are used for the analysis. A discussion on the reliability of the use of energy distribution and Relative Wavelet Energy calculations for distinguishing unipolar depression patients from healthy controls is presented. The potential of the application of Wavelet Entropy to differentiate states of the brain under normal and pathologic condition is introduced. Details are given on the suitability of ascertaining certain nonlinear indices on the feature extraction, assuming the time series to be highly nonlinear. The assumption of nonlinearity of the measured EEG time series is further verified using surrogate analysis. The studies discussed in this chapter indicate lower values of nonlinear measures for patients. The higher values of signal energy associated with the delta bands of depression patients in the lower frequency range are regarded as a major characteristic indicative of a state of depression. The chapter concludes by presenting the important results in this direction that may lead to better insight on the brain activity and cognitive processes. These measures are hence posited to be potential biomarkers for the detection of depression.

Keywords: depression, relative wavelet energy, wavelet entropy, approximate entropy, Hurst exponent, largest Lyapunov exponent and fractal dimension, surrogate analysis

1. Introduction

Depression refers to a state of mental disorder accompanied by mood variations that affect the thought process, social and physical well-being of an individual. World Health Organisation reports that more than 264 million people under all age group suffer globally from this leading cause of disability. Depression may also sometimes lead to cognitive impairment. History demonstrates that depressive disorders have been with human civilization from the very beginning of the mankind. Unlike many other ailments that affect the health of an individual, an early diagnosis of this mental disorder is highly challenging. Timely medical intervention has been proved to be very effective in arresting the progression of this disorder. Automated diagnosis using EEG signals of the brain would be highly beneficial in the effective clinical intervention and thereby assisting the psychiatrists in the assessment of mental state.

EEG signals contain information about the state of the brain. The variations in the biosignals, indicating certain symptoms, are highly subjective and may appear at random in time scale. The electroencephalogram has been used as a tool for investigating the brain electrical activity in different physiological and pathological states for several decades. The identification of neurophysiological events, different behavioural states and the localisation of the areas involved constitute a relevant task in the EEG analysis.

Electroencephalogram (EEG) is the recording of electrical activity along the scalp produced by the firing of neurons within the brain. It is a tool which helps in diagnosing various disorders of the brain and also helps in studying the functional state of the brain. EEG recording is most commonly done by placing the electrodes on the scalp while localised measurement of potentials is done subdurally or from the cerebral cortex. Electrode placement for recording EEG is based on the International 10–20 electrode placement system. The amplitude of the EEG signal is slightly less than 10 μV to slightly more than 100 μV p–p and the frequency ranges from 0 to slightly greater than 100 Hz. Earlier, the analysis of EEG has been based on the assumption that the EEG signals are generated by a highly complex linear system, but later they have been interpreted as the output of a deterministic system of relatively low complexity but containing nonlinear elements. Thus applying the concept of deterministic chaos to the EEG, it can be characterised by various parameters [1]. EEG studies of depression patients have been proven worthwhile for quantitative analysis that will lead to the development of automated clinical diagnostic tools.

In this chapter we discuss the method to characterise and compare frontopolar-temporal EEG signals of depression patients and normal controls using signal processing and nonlinear methods. An 8-level Multiresolution decomposition of the time -frequency analysis, which decomposes a mixed signal into signals at different frequency bands, is attempted. Energy at different resolution levels has been calculated using Parseval's theorem. Relative Wavelet Energy (RWE) is used to characterise the EEG signal energy distribution of healthy subjects and depression patients at different frequency bands. Wavelet Entropy calculations are performed to assess the degree of order associated with the acquired signals. All the aforementioned measures posit better quantitative measures in the comparative study of brain activity and complexity in depression patients and normal controls.

2. Materials and methods

2.1 Measurement protocol

The real time data was recorded from 30 medication free outpatients under the age group of 20–50 years comprising of 16 female and 14 male patients from the Psychiatry department of Medical College, Calicut, Kerala, India (female mean age: 33; male mean age: 35). The measurement was done on unipolar depression patients who did not have any history of substance abuse and no significant medical illness.

Similarly 30 age and sex matched healthy controls also participated in the study who were free of medical illness. None of them reported of a history of any central nervous system disorder. Bipolar EEG recordings using a 24-channel EEG measuring instrument was carried out at locations FP1-T3 (left half) and FP2-T4 (right half) of the brain. The electrodes were placed, based on the International 10–20 electrode placement system. An ear clip electrode attached to the right earlobe served as an isoground connection. The EEG recordings were done by placing the electrodes on the frontopolar-temporal regions both on the left and right half of the brain for duration of 5 minutes each, under eyes closed and eyes open condition in a resting state. The sampling frequency of the signal is 256 Hz and is notch filtered at 50 Hz to remove the power line interference. Statistical analysis was performed by One-way ANOVA to test for differences among the two classes of EEG signals recorded. Informed written consent was obtained from all the subjects who participated in the study and medical ethical committee approval was taken prior to the study. **Figures 1** and **2** show a typical EEG signal of normal and depression patient respectively.

2.2 Preprocessing

Artefacts such as eye movements, eye blinks, head movements, cardiac and muscle activation artefacts, tongue movements and power line noise pose a problem for the proper EEG interpretation and analysis. Other artefacts that disrupt the EEG signal include instrument artefacts (faulty electrodes), sweat artefacts, impedance fluctuations, cable movements, pulse artefacts etc. Power line interferences are removed from the EEG signal by using a 50 Hz notch filter. Eye movement and muscle movement artefacts are manually removed from the signal with the help of an expert by visual inspection. In this work, the high frequency components present in the acquired EEG signals are denoised using Total Variation Filtering (TVF) [2].

The TVF employed in this work is based on the algorithm developed by Chambolle [3]. A dual formulation approach is used to minimise the objective

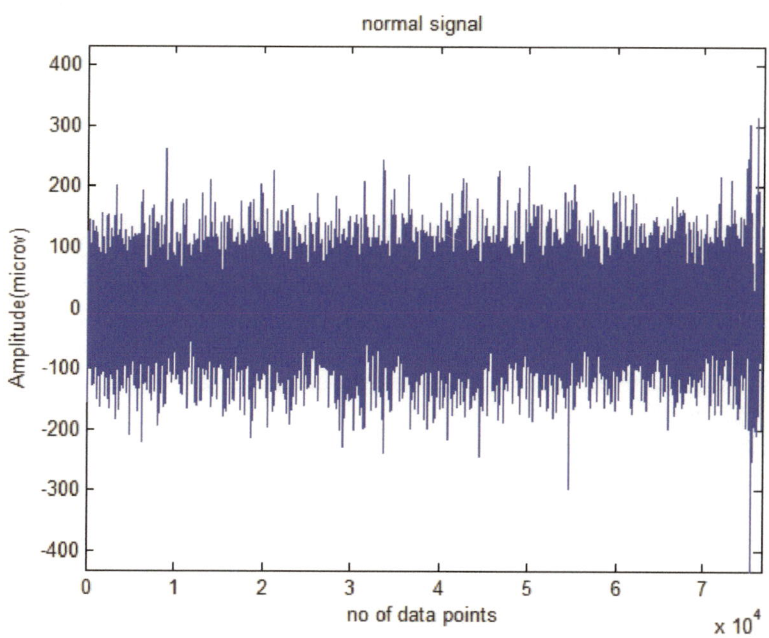

Figure 1.
EEG signal of a normal control.

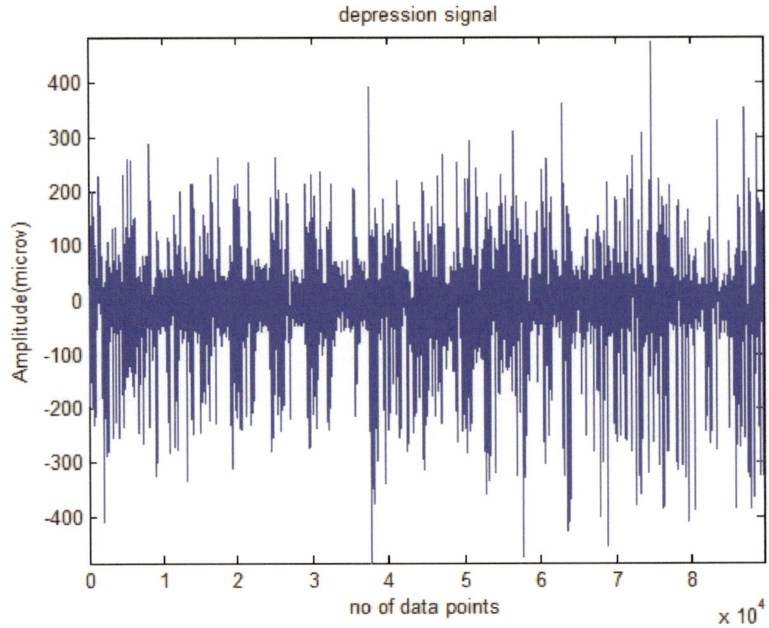

Figure 2.
EEG signal of a depression patient.

function of the Total Variation (TV) denoising problem. So the TV denoising problem amounts to minimising the following discrete function [4]:

$$J(x) = \|y - x\|_2^2 + \lambda \|Ax\|_1 \tag{1}$$

where A is a matrix of size M × N. Smoothing of the signal is controlled by λ, which is known as the regularisation parameter. Since the amount of high frequency noise present in the EEG signal recorded from depression patients using the 24-channel equipment was already low, the optimal value of λ for denoising was found to be 0.9.

2.3 Selection of wavelet

From an array of discrete orthogonal wavelets, Daubechies-1(D1) to D10, Coiflet-1 to Coiflet- 5 and Symlet 1 to Symlet 8, the task is to identify the wavelet which suits well with the individual EEG signal recorded from depression patients. This is necessitated as it is found that there is extreme patient variability and also variability of the signals with respect to the location on the scalp from person to person. All the above 23 wavelets were tested on all the 30 patient records under four categories namely eyes open and eyes closed conditions recorded from the left and right half of the brain.

The best wavelet is chosen based on the highest value of correlation coefficient which indicates a better match of the characteristics of the EEG signal of depression patient with the wavelet selected. Of the 30 cases considered in this experimental study described here, for 85% of the cases, Coiflet 5 emerged as the best suited wavelet. Hence Coiflet 5 is used for analysis.

2.4 Wavelet Transforms

Wavelet Transforms are efficiently used in many of the signal processing applications as it gives more accurate time and frequency representation of the signal.

Wavelet Transforms help in the extraction of wavelet coefficients of discrete time signals. An important feature of Wavelet Transform is that it gives good frequency resolution over a large window while good time resolution at high frequencies. This feature has been of great interest to biomedical applications, as most of the biosignals contain high frequency components in a short span and low frequency components over large span. Wavelet Transforms are thus highly useful for the analysis of nonlinear, nonstationary signals as it gives an excellent time–frequency resolution.

In this method, the signal is decomposed into a set of basis functions called wavelets by the Wavelet Transform. These basic functions are obtained by dilations, contractions and shifts of a unique function called wavelet prototype. Continuous wavelets are functions generated from one single function by dilations and translations of a unique mother wavelet $\psi\ (t)$:

$$\Psi_{a,b}(t) = \frac{1}{\sqrt{|a|}}\Psi\left(\frac{t-b}{a}\right) \tag{2}$$

where a is the scale parameter, b, the shifting parameter and t, the time. The function set $(\Psi_{a,b}(t))$ is called the wavelet family. The Wavelet Transform usually used in engineering application is the Discrete Wavelet Transform (DWT). It uses the discrete values of the scaling and the translational parameters given by, $a = a_0^j$ and $b = kb_0a_0^j$ where j and k are integers. Then we get:

$$\Psi_{j,k}(t) = a_0^{-j/2}\Psi\left(a_0^{-j}t - kb_0\right) \tag{3}$$

where j indicates frequency localization and k indicates time localization. Dyadic scheme implementation is the basis for Multiresolution Analysis (MRA) in Discrete Wavelet Transforms. Any time series can be decomposed in terms of coarse approximations provided by scaling functions and the detail information by the wavelet functions [5]. The scaling function is associated with low-pass filters (LPF) and the wavelet function is associated with the high pass filters (HPF). The decomposition of the signal into the different frequency bands is obtained by successive convolution with high-pass and low pass filtering of the time domain signal.

The approximations are the low frequency components and the details are the high frequency components of the time series. The detail coefficients and approximation coefficients at level 1 (CD1 and CA1) are obtained by decimating the outputs from both the filters by 2. The procedure is then repeated by sending the approximation coefficients to the second stage. This is continued till the signal is decomposed at the expected level. In this work, an eight level decomposition is carried out. The EEG signal acquired from the depression patients are sampled at a frequency of 256 Hz. The multiresolution decomposition offers a time-frequency decomposition of the signal involving not only its energy but also the morphological aspects that are relevant for signal recognition and understanding [6]. Each of the wavelet scales corresponds to a specific frequency band given by

$$f = \frac{2^{n-m}f_s}{2^n} \tag{4}$$

where, f is higher frequency limit of the frequency band represented by decomposition level m, f_s is sampling frequency and 2^n is the number of data points in the signal. CD1 (64-128 Hz) and CD2 (32-64 Hz) correspond to gamma band, CD3 (16-32 Hz) corresponds to beta band, CD4 (8-16 Hz) corresponds to alpha band, CD5 (4-8 Hz) corresponds to theta band while CD6 (2-4 Hz), CD7 (1-2 Hz), CD8

(0.5-1 Hz) and CA8 (0–0.5 Hz) correspond to delta band. The DC level of the signal corresponding to eighth level approximation coefficient (CA8) is also considered in this work. The signals are reconstructed from the wavelet coefficients for each scale by applying inverse transform. The reconstructed signal coefficients obtained from eight levels of details and from the eighth level of approximation are further used for energy calculations.

2.5 Calculation of energy

After multilevel decomposition, Parseval's theorem is employed to calculate the energy of the reconstructed signal coefficients at the detail and approximation levels. It helps in identifying the energy distribution in different frequency bands of gamma, beta, alpha, theta and delta [7]. Parseval's theorem is mathematically expressed as [8]:

$$\sum_{n=1}^{N} |x(n)|^2 = \sum_{n=1}^{N} |a_j(n)|^2 + \sum_{j=1}^{m} \sum_{n=1}^{N} |d_j(n)|^2 \tag{5}$$

where $x(n)$ is the time domain discrete signal, N is the total number of samples in the signal, $\sum_{n=1}^{N}|x(n)|^2$ is the total wave energy of the signal $x(n)$, $\sum_{n=1}^{N}|a_j(n)|^2$ is the total energy concentrated in the level 'j' of the approximated version of the signal. $\sum_{j=1}^{m}\sum_{n=1}^{N}|d_j(n)|^2$ is the total energy concentrated in the detail version of the signal, from level 1 to m and m is the maximum level of wavelet decomposition.

Energy distribution calculations are done on all recordings of eyes open and eyes closed conditions, acquired from both the left (FP1-T3) and right (FP2-T4) frontopolar-temporal regions of the brain for both the normal controls and depression patients. Energy distribution associated with the various bands of frequency, for different levels of detail are plotted in **Figures 3–7**, for a single case of measurement recorded (from the left half of the brain under eyes closed condition). Similar variations are also observed for measurements with the other recording protocols.

It is observed that there exists a clear difference in the energy levels of EEG signals of both normal and depression patients in the Gamma band (D1) (**Figure 3**). Values of energy obtained for normal controls are always higher than that of depression patients, covering all age groups. These differences in energy levels tend to narrow down as we move from Gamma band D1 to alpha band D4 through D2 and D3. It is interesting to note that normal subjects register higher energy distribution levels for cases considered up to theta band D5. This trend appears to get

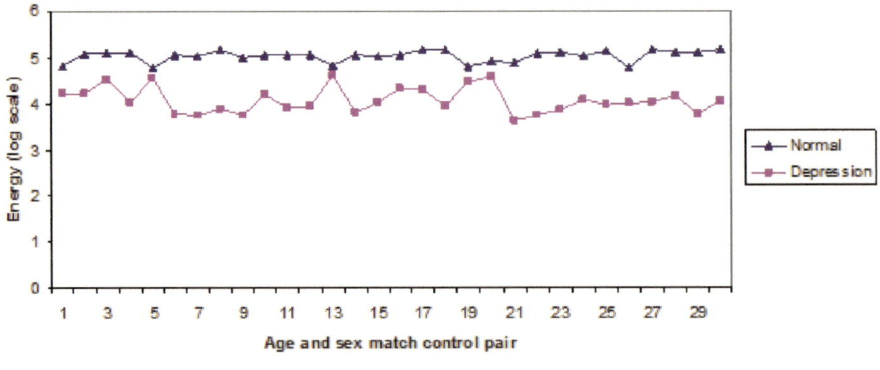

Figure 3.
Energy distribution in gamma band (D1) of normal controls and depression patients.

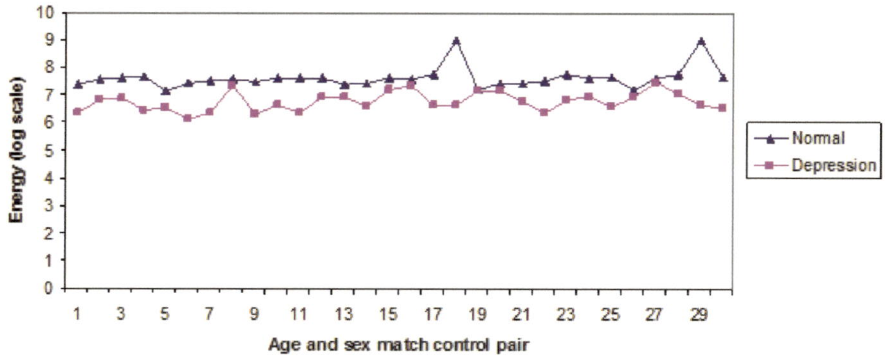

Figure 4.
Energy distribution in Beta band (D3) of normal controls and depression patients.

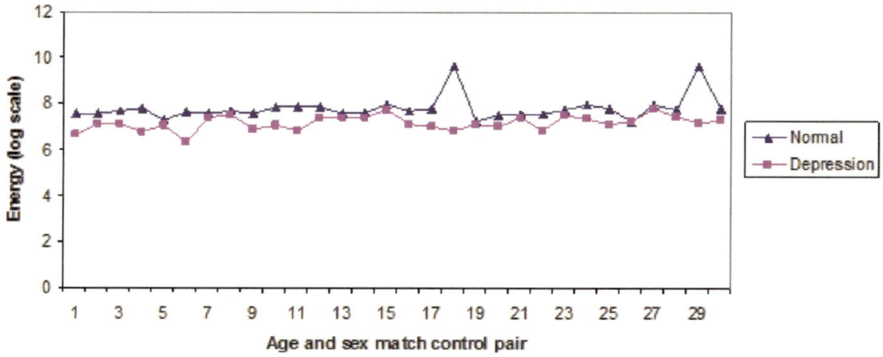

Figure 5.
Energy distribution in alpha band (D4) of normal controls and depression patients.

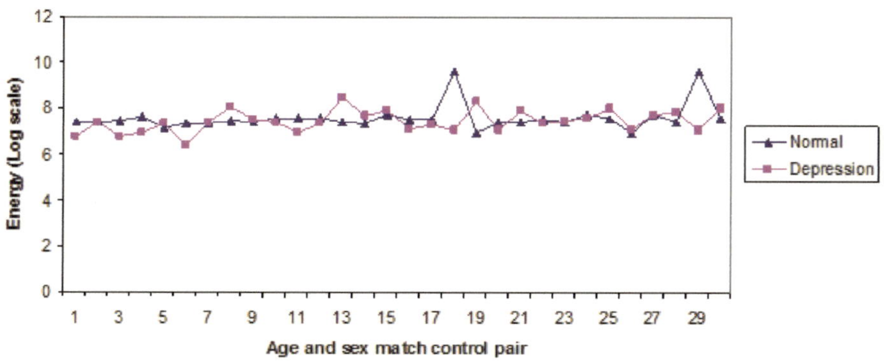

Figure 6.
Energy distribution in theta band (D5) of normal controls and depression patients.

reversed beyond theta band D5 (see **Figure** 7 for Delta band D8). The energy distribution of depression patients in theta band (D5-**Figure 6**) is almost at par with that of normal controls. But the energy distribution of depression patients has crossed the threshold of normal subjects and is higher than normal in all cases of delta band from D6-D8 and A8. It gives a clear indication that the brain activity of depression patients in gamma, beta and alpha band is lower when compared to healthy controls. Similar is the trend for all other measurements taken from the right side of the brain under eyes closed and open condition and also for measurements from the left half of the brain for eyes open condition.

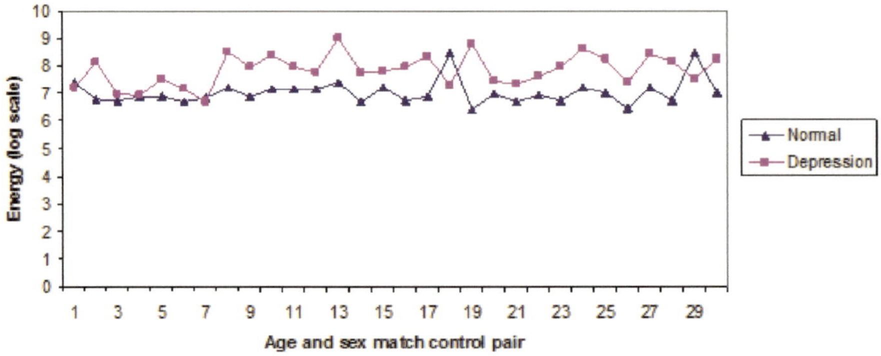

Figure 7.
Energy distribution in delta band (D8) of normal controls and depression patients.

2.6 Relative Wavelet Energy (RWE)

Relative Wavelet Energy (RWE) gives information about the relative energy associated with the frequency bands and it can detect the degree of similarity between segments of a signal [9, 10]. It is also known from previous studies that RWE is a good tool for detecting and characterising specific phenomenon in time and frequency planes [11].

The energy wavelet coefficients $d_{j,k}$ represent the detail signal energy at each resolution level, j given by,

$$E_j = \sum_k |d_{j,k}|^2 \quad j = 1 \ldots N \tag{6}$$

The energy scaling coefficients C_k is defined as the energy at decomposition level $N + 1$ given by

$$E_{N+1} = \sum_k |c_k|^2 \tag{7}$$

Thus the total energy of the signal for all levels is given by

$$E_{total} = \sum_{j=1}^{N+1} E_j \tag{8}$$

and hence the Relative Wavelet Energy (RWE) is defined as

$$\rho_j = \frac{E_j}{E_{total}} \; j = 1, \ldots, N+1 \tag{9}$$

Clearly $\sum_j \rho_j = 1$ and the distribution $\{\rho_j\}$ can be considered as a time scale density. This provides information to characterise signal energy distribution at different frequency bands.

RWE calculations are carried out on the different frequency bands to understand the variations in healthy subjects and depression patients. Wavelet energy in the gamma band, particularly D1 is negligible, while D2 is negligibly small for depression patients in comparison to normal controls. The RWE levels associated with beta (D3–18.5%) and alpha (D4–22%) bands for normal controls are approximately 25% higher than the corresponding values for depression patients with D3 being 1.9% and D4–4.3% (**Figures 8** and **9**). RWE of theta band also shows similar trend

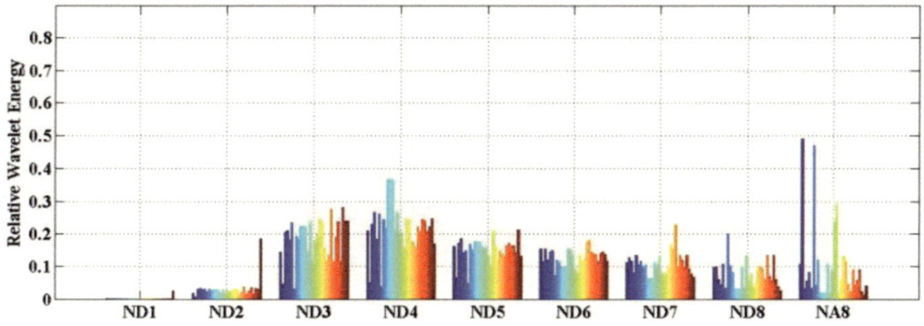

Figure 8.
RWE of all frequency bands of normal controls from the left half of the brain under eyes open condition.

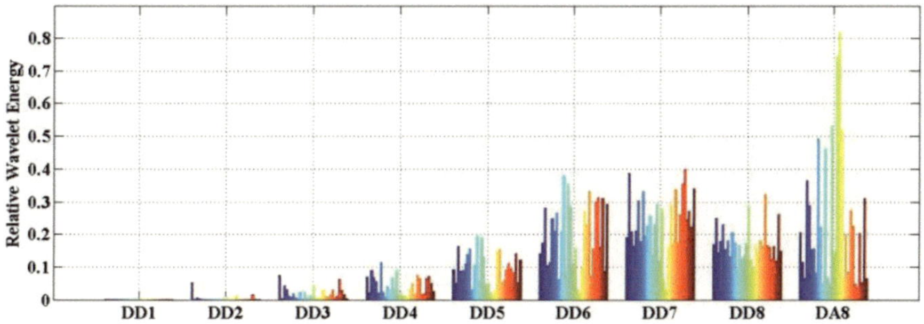

Figure 9.
RWE of all frequency bands of depression patients from the left half of the brain under eyes open condition.

with normal controls having values slightly lower than twice the values associated with depression patients. RWE associated with delta band (D6–19.5%, D7–23.5%, D8–17.1% and A8–23.7%) show appreciably higher values for depression patients when compared to normal controls. The feature worth noting is that RWE of approximation level has the highest percentage in depression patients to that of normal controls. From the results of RWE calculations, it may be observed that the RWE is more prominent in depression patients in the frequency bands, 1-2 Hz and 0–0.5 Hz. Hence depression may be classified as a very low frequency phenomenon.

RWE values plotted in **Figures 8** and **9** represent the calculations carried out on the frequency bands of normal and patient EEG signals recorded from the left half of the brain under eyes open condition. RWE calculations are also carried out on the frequency bands of the EEG signal acquired from the left half of the brain under eyes closed condition and on the EEG signals from the right half of the brain both under eyes open and eyes closed conditions. The observation from all the protocols reveals a high value of RWE in alpha band (D4) of normal controls indicating high activity in the thought process of healthy subjects. Also a high value of RWE in 8th level approximation followed by detail level D7 is observed in depression patients. Hence from all the four cases, it may be concluded that depression phenomenon is confined to the lower frequency bands especially in delta band of 0–4 Hz. In order to analyse the statistical nature of the measurement among the two broad classes of EEG signals recorded, One-way ANOVA is carried out which gave a statistical significant difference ($p < 0.005$).

2.7 Wavelet Entropy (WE)

The degree of order/disorder associated with a multifrequency signal response is characterised by Wavelet Entropy (WE). The time evolution of WE was also calculated which gave information about the dynamics in the EEG records. It was observed that in contrast to spectral entropy, WE is capable of detecting changes in a nonstationary signal due to the localisation characteristics of the Wavelet Transform. The computational time of WE was significantly shorter since the algorithm involved the use of fast wavelet transform in a multiresolution framework. The results demonstrated that WE could differentiate between specific physiological brain states under spontaneous or stimulus-related conditions. The time evolution of the WE is calculated to give information about the dynamics in the EEG records. A signal generated by a totally random process can be taken as representative of a very disordered behaviour.

The Shannon Wavelet Entropy (WE) as a function of time is calculated [10] as:

$$WE = -\sum_{j=1}^{m} \rho_j \ln\left(\rho_j\right) \tag{10}$$

where m is the wavelet decomposition level from level 1 to level m.

WE may be considered as a meaningful indicator since it is able to differentiate physiological brain states under normal and depression conditions. **Figure 10** represents the Wavelet Entropy calculated for the EEG signals recorded from the left half of the brain under eyes open condition from normal and depression patients. Significant decrease in the WE is observed in the EEG signals of depression patients under all the four recording protocols, indicating a more rhythmic and ordered behaviour of the EEG signal [12, 13]. Being independent of the amplitude or the energy of the signal, the WE gives additional information about EEG signals in comparison to those obtained by using frequency analysis or other standard methods. The use of such quantifiers based on time-frequency methods can contribute to the analysis of brain responses and may also lead to a better understanding of their dynamics.

2.8 Nonlinear measures

Reduction in complexity in patients with disease is the main hypothesis that is checked in most of the research work. Here we analyse the EEG signal complexity and irregularity of the frontopolar-temporal regions of the brain of controls and

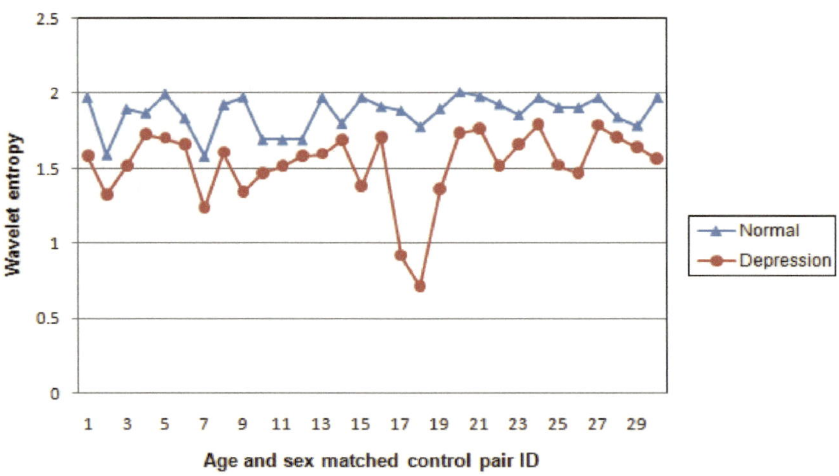

Figure 10.
Wavelet entropy calculated for the EEG signals recorded from the left half of the brain under eyes open condition from normal and depression patients.

patients with unipolar depression under resting states of eyes open and closed conditions. The analysis is carried out using Approximate Entropy, Fractal Dimension and Largest Lyapunov exponent.

2.8.1 Approximate Entropy (ApEn)

Approximate Entropy (ApEn) is a statistic quantifying regularity and complexity which has potential application to a wide variety of physiological and clinical time series data. ApEn is a statistical parameter that measures the predictability of the current amplitude values of a physiological signal based on its previous amplitude values. Approximate Entropy is the is the probability difference of the pattern similarities of the connected straight lines of the m adjacent points and $m + 1$ adjacent points of time sequence data [14]. The more complex the sequence data, higher is the probability that new pattern appears and larger the corresponding ApEn. ApEn measures the (logarithmic) likelihood that runs of patterns that are close remain close on next incremental comparisons.

For a time series of N data points, $\{u(i):1 \leq i \leq N\}$, form vector sequences $x(1)$ through $x(N\text{-}m + 1)$, defined by $x(i) = [u(i),..., u(i + m\text{-}1)]$. These vectors represent m consecutive u values, commencing with the ith point. m is the length of compared runs. The distance $d[x(i), x(j)]$ between vectors $x(i)$ and $x(j)$ is defined as the maximum difference in their respective scalar components. Let B_i be the number of vectors $x(j)$ within r of $x(i)$ for a window length m and let A_i be the number of vectors $x(j)$ within r of $x(i)$ for a window length $m + 1$, where r is the tolerance for accepting matches. The function $C_i^m(r)$ is defined as:

$$C_i^m(r) = \frac{(B_i)}{(N - m + 1)} \tag{11}$$

In calculating $C_i^m(r)$, the vector $x(i)$ is called the template and the instance where a vector $x(j)$ is within r of it is called a template match. $C_i^m(r)$ is the probability that any vector $x(j)$ is within r of $x(i)$. It measures within a tolerance r, ($r = k$*standard deviation*)* the regularity, or frequency, of patterns similar to a given pattern of window length m. The function $\Phi^m(r)$ is defined as:

$$\Phi^m(r) = \frac{1}{N - m + 1} \sum_{i=1}^{N-m+1} \ln C_i^m(r) \tag{12}$$

where, $\Phi^m(r)$ is the average of the natural logarithms of the functions $C_i^m(r)$. For finite data sets,

$$ApEn(m, r, N) = \Phi^m(r) - \Phi^{m+1}(r) \tag{13}$$

The parametres N, m and r must be fixed for each calculation and r effectively works as a filter. The values of m and k adopted in this work are 1 and 0.2 respectively.

The column plot of **Figure 11** represents the ApEn values calculated for the EEG signals of healthy controls and patients, acquired from the left half of the brain under eyes open condition. The results of ApEn calculated for from the left part of the brain under eyes closed conditions and from the right part of the brain both under eyes open and closed conditions indicate that normal controls have a higher value of ApEn than depression patients. A low value of ApEn indicates predictability and regularity in a time series, whereas a high value of ApEn indicates

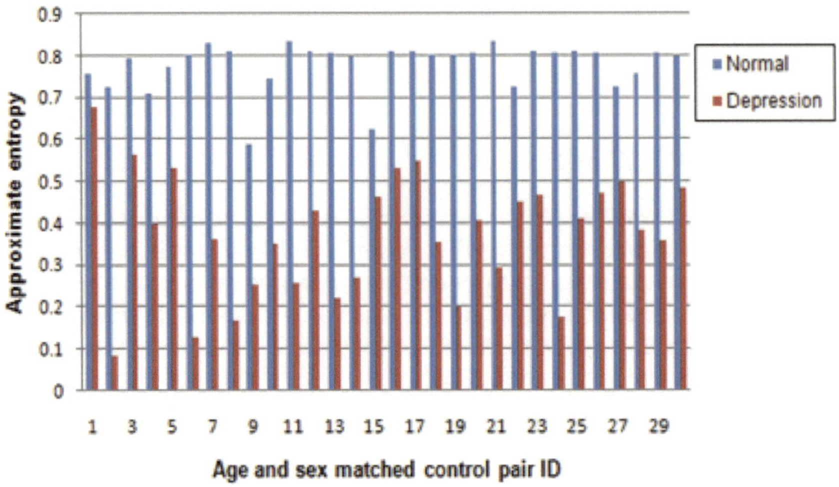

Figure 11.
Approximate entropy values calculated for the EEG signals of normal and depression patients acquired from the left part of the brain under eyes open condition.

unpredictable and random variation. The results of ApEn indicate that the complexity of the brain is high in normal controls while the signals from depression patients are less complex.

2.8.2 Fractal Dimension (FD)

FD analysis is frequently used in biomedical signal processing, including EEG analysis. Fractal Dimension can be used to quantify the complexity and the self-similarity of an object. The EEG FD can expected to be always between 1 and 2 since the dimension of a plane is equal to 2 and the dimension of a line is equal to 1. FD can be calculated using Higuchi's method, Katz algorithm, box counting approach and so on. FD analysis provides a fast computational tool to track complexity variations of biosignals. FD analysis used in this work is based on Higuchi's algorithm [15]. The algorithm is based on the measure of the mean length of the curve $L(k)$ by using a segment of k samples as a unit of measure.

Consider $x(1), x(2), \ldots, x(N)$ be the time series to be analysed. The algorithm constructs k new time series x_m^k, defined as

$$x_m^k : x(m), x(m+k), x(m+2k), \ldots\ldots\ldots, x\left(m + \lfloor \frac{N-m}{k} \rfloor k\right) \, for \; m = 1, 2, \ldots., k \quad (14)$$

where m and k are integers indicating the initial time and the interval time respectively. $\lfloor a \rfloor$ means the integer part of a. For each of the curves or time series x_m^k constructed, the average length $L_m(k)$ is computed as

$$L_m(k) = \frac{\sum_{i=1}^{\lfloor N-m/_k \rfloor} |x(m+ik) - x(m+(i-1)k)|(N-1)}{\lfloor \frac{N-m}{k} \rfloor k} \quad (15)$$

where N is the total length of the data sequence x and $\frac{(N-1)}{\lfloor (N-m)/_k \rfloor k}$ is a normalisation factor.

An average length is computed for all time series having the same delay (or scale) k, as the mean of the k lengths $L_m(k)$ for m = 1,….,k. This procedure is

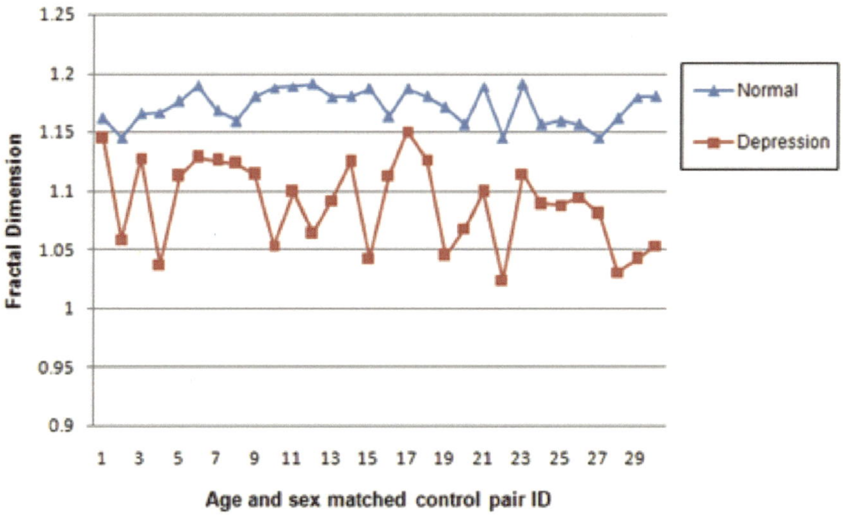

Figure 12.
Fractal dimension values of EEG signals of normal and depression patient from the left part of the brain under eyes closed condition.

repeated for each k ranging from 1 to k_{max}, yielding a sum of average lengths $L(k)$ for each k as indicated in Eq. (16).

$$L(k) = \sum_{m=1}^{k} L_m(k) \tag{16}$$

The total average length for scale k, $L(k)$ is proportional to k^{-D}, where D is the FD by Higuchi's method. In the curve of $\ln(L(k))$ versus $\ln(1/k)$, the slope of the least squares linear best fit is the estimate of the FD. The k_{max} values depends on the dimension D, on the signal's length N and on the specific class of fractal signals. The value of k_{max} chosen for the analysis of EEG signals is 6. **Figure 12** shows the plot of Fractal Dimension values calculated from the EEG signals of normal and depression patients acquired from the left part of the brain under eyes closed condition.

FD based on the algorithm followed is a quantifier evaluated directly in the time domain. Similar to the plot in **Figure 12**, the plots of FD for the rest of the recording protocols show that the values of FD are higher for normal controls indicating higher complexity in EEG signals of normal controls [16, 17]. It may be concluded that the value of FD increases with the increase in the degree of the cognitive activity. Lower values of FD indicate a low degree of cognitive activity for depression patients.

2.8.3 Largest Lyapunov Exponent (LLE)

Lyapunov exponents provide a qualitative and quantitative characterisation of dynamical behaviour. To discriminate between chaotic dynamics and periodic signals, Largest Lyapunov Exponent (λ) is often used. Lyapunov exponents are the average exponential rates of divergence or convergence of nearby orbits in phase space. It is a quantitative measure of the sensitive dependence on the initial conditions. The rate of separation between the nearby orbits in phase space is characterised by the Largest Lyapunov Exponent λ, mathematically written as:

$$\|X_\tau - Y_\tau\| \approx \|X_0 - Y_0\| e^{\lambda\tau} \tag{17}$$

where X_0 and Y_0 are two initial conditions close together, and X_τ and Y_τ are their respective time evolutions after τ time units. The approach of Wolf et al. [18] was used for constructing the algorithm for calculating the Largest Lyapunov Exponent (LLE) in this work.

Let $\overrightarrow{x_0}(t)$ denote a reference trajectory passing through $\overrightarrow{x_0}(0)$ at time $t = 0$ and let $\overrightarrow{x_1}(t)$ denote a trajectory passing through $\overrightarrow{x_1}(0)$ at time $t = 0$. The Largest Lyapunov Exponent $\lambda(\overrightarrow{x_0})$ is defined with respect to the reference orbit $\overrightarrow{x_0}$ by

$$\lambda(\overrightarrow{x_0}) = \lim_{t\to\infty} \lim_{\left\|\overrightarrow{\Delta x(0)}\right\|\to 0} \frac{1}{t} \log \frac{\left\|\overrightarrow{\Delta x(t)}\right\|}{\left\|\overrightarrow{\Delta x(0)}\right\|} \tag{18}$$

where $\left\|\overrightarrow{\Delta x(0)}\right\|$ is the Euclidean distance between the trajectories $\overrightarrow{x_0}(t)$ and $\overrightarrow{x_1}(t)$ at an initial time $t = 0$ and $\left\|\overrightarrow{\Delta x(t)}\right\|$ is the Euclidean distance between the trajectories $\overrightarrow{x_0}(t)$ and $\overrightarrow{x_1}(t)$ at a later time t. In this definition $\overrightarrow{x_1}(t)$ can be any trajectory that is initially infinitesimally close to $\overrightarrow{x_0}(0)$ at time $t = 0$. The correspondence between sensitivity to initial conditions and a positive Lyapunov exponent is obvious in equation number 17. An embedding dimension of 10 and a delay of 1 were used for calculating LLE. **Figure 13** shows the values of LLE calculated for normal and depression patients from the left half of the brain under eyes closed condition.

Larger values of LLE observed for all the recording protocols for normal controls are indicative of higher brain activity [16, 17]. Therefore LLE can be effectively used for discriminating the EEG signals of normal controls and depression patients.

2.9 Surrogate data analysis

The method of using surrogate data in time-series analysis was introduced by Theiler et al. to validate that a given time-series is nonlinear. Nonlinear indices such

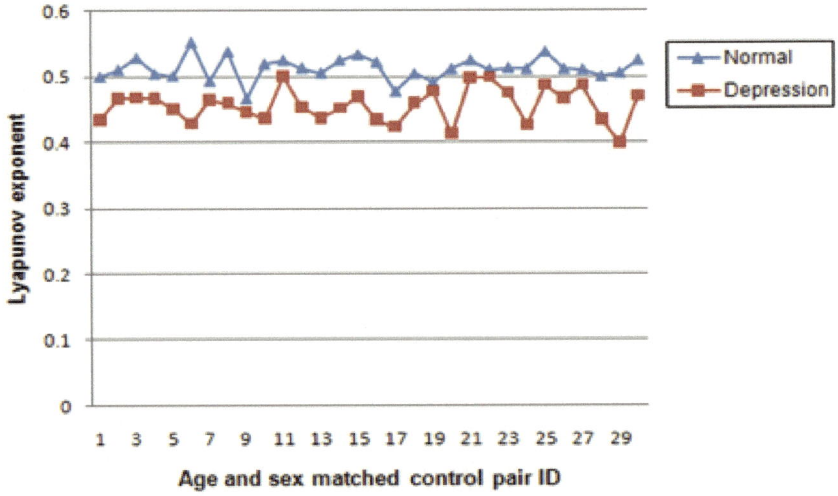

Figure 13.
LLE values of EEG signals acquired from normal controls and depression patients from the left part of the brain under eyes closed condition.

as FD *and* LLE are computed for several surrogate data series. Their values are compared with that of the nonlinear index computed for the original data. The lack of any statistically significant difference is interpreted as the deviation from a linear process. Surrogate data is constructed by phase randomising the original time series and has the same linear features like mean, variance, histogram and power spectrum as the original data. This method of generating surrogate data is based on the amplitude adjusted Fourier transform method, which yields the same distribution of amplitudes but randomises the phases from the spectral aspect [19]. Tentative surrogate data are obtained by inverse Fourier transform.

To test for a statistical significance of difference in *FD* and *LLE* between the original and the surrogate data, 10 surrogate data series were generated to match each original signal. Let $LLE\ (D)$ be the *LLE* of the original data, and let $LLE\ (Si)$ be the *LLE* of the 10 surrogate series ($i = 1, \ \dots, 10$). The mean and standard deviation, SD of $LLE\ (Si)$ ($i = 1, \ \dots, 10$) are estimated as $LLE\ (S)$ and $SD(LLE(Si))$. The statistical significance measure σ then is computed as follows:

$$\sigma = \frac{|LLE(D) - LLE(S)|}{SD(LLE(Si))} \tag{19}$$

It follows a Student *t* test distribution with 9 degrees of freedom ($\sim t_{9[1-\alpha/2]}$). For $\alpha = 0.05$, the critical value of *t* is 2.26. Accordingly, when the $\sigma > 2.26$, the null hypothesis is rejected at the 5% probability level, and the original data are considered to contain nonlinear features.

Nonlinear indices (FD and LLE) are computed for several surrogate data series and their values are compared with the ones computed for the original series. The demonstration of significant difference in nonlinear indices between the original and surrogate data is supportive of the presence of nonlinearity in the original data. **Tables 1** and **2** show the calculation of the statistical significance measure for both the normal controls and depression cases for the nonlinear indices, FD and LLE. The results prove that the original data contain nonlinear features since the statistical significance measure is greater than 2.26.

Protocol	Fractal dimension			Largest Lyapunov exp.		
	Original data	Surr. data	Stat. sig. Meas.	Original data	Surr. data	Stat. sig. Meas.
L:EC	1.1659	1.1889	**65.22**	0.552	0.5059	**3.87**
L:EO	1.159	1.1803	**50.14**	0.538	0.6066	**2.84**
R:EC	1.1663	1.1875	**63.75**	0.529	0.6177	**5.25**
R:EO	1.6556	1.1876	**76.04**	0.485	0.5251	**3.85**

Table 1.
Surrogate data analysis of normal controls.

Protocol	Fractal dimension			Largest Lyapunov exp.		
	Original data	Surr. data	Stat. sig. Meas.	Original data	Surr. data	Stat. sig. Meas.
L:EC	1.0368	1.0604	**57.07**	0.3978	0.4808	**7.36**
L:EO	1.0441	1.0615	**35.63**	0.4811	0.5201	**2.45**
R:EC	1.1220	1.1536	**62.23**	0.3767	0.4782	**4.43**
R:EO	1.0638	1.0827	**47.95**	0.5420	0.4775	**4.93**

Table 2.
Surrogate data analysis of depression patients.

3. Conclusions

The characteristics of frontopolar-temporal EEG signals of depression patients are investigated using signal processing techniques and nonlinear parameters. EEG signals for the analysis were acquired from 30 unipolar depression patients and 30 age and sex matched healthy controls. Bipolar EEG recording using a 24-channel EEG machine was carried out at locations FP1-T3 (left half) and FP2-T4 (right half) of the brain for a duration of 5 minutes each, under eyes closed and eyes open condition in a resting state. Total variation filtering was found to be effective in removing the high frequency noise while the eye blink and eye movement artefacts were removed by visual inspection. Wavelet analysis is performed and signal features having significant influence on the signal waveforms of depression and normal controls have been identified. Coiflet 5 is used for the wavelet analysis. An 8-level decomposition was carried out and Relative Wavelet Energy was calculated on the reconstructed signal coefficients. Wavelet Entropy calculations revealed the degree of disorder associated with the EEG signals. The nonlinear measures like Approximate Entropy (ApEn), Fractal Dimension (FD) and Largest Lyapunov Exponent (LLE) are calculated. Nonlinearity of the EEG signal under study was confirmed by surrogate data analysis.

Depression effects are reflected mainly in the lower frequency range indicating a reduced brain activity. The multiresolution decomposition characterised the various frequency bands of EEG signals. The wavelet energy distribution in different frequency bands indicated higher levels of brain activity for normal controls in gamma, beta and alpha bands. It also showed lower brain activity in the delta bands of depression patients. The results from the calculations of RWE confirm the fact that, mental activity as reflected in the EEG signals of depression patients is confined to the lower frequency range especially in the delta band of 0-4 Hz. The quantitative evaluation of nonlinear parameters like ApEn, FD and LLE confirmed higher brain activity for normal controls compared to depression patients. Lower values of these nonlinear parameters indicate the fact that complexity of EEG is reduced in depression which effectively helped in discriminating EEG signals of depression patients and healthy controls. The quantitative assessment of signal characteristics and nonlinear parameters from the present study may be of significant use in the analysis of brain dynamics.

Author details

Subha D. Puthankattil
Electrical Engineering Department, National Institute of Technology Calicut, Kozhikode, Kerala, India

*Address all correspondence to: subhadp@nitc.ac.in

References

[1] Subha DP, Paul KJ, Rajendra UA, Lim CM. EEG signal analysis: A survey. Journal of Medical Systems. 2010;**34**: 195-212. DOI: 10.1007/s10916-008-9231-z

[2] Rudin L, Osher S, Fatemi E. Nonlinear total variation based noise removal algorithms. D. 1992;**60**:259-268. DOI: 10.1016/0167-2789(92)90242-F

[3] Chambolle A. Total variation minimization and a class of binary MRF models. In: Proceedings of the 5th international conference on energy minimisation methods in computer vision and pattern recognition (EMMCVPR 2005). November. 2005; **37357**:136-152. DOI: 10.1007/11585978_10

[4] Ivan WS, Ilker B. Total Variation Filtering [Internet]. Available from h ttp://eweb.poly.edu/iselesni/teaching/lecture_notes/TV_**filtering**.pdf. *4th February* 2010.

[5] Mallat S. A theory for multiresolution signal decomposition: The wavelet representation. IEEE Trans. Pattern Anal. Mach.Intell. 1989;**11**:674-693. DOI: 10.1109/34.192463

[6] Ismael C, Rolando B, Maribel E, Trinidad V. Multiresolution decomposition of nonstationary EEG signals: A preliminary study. Computers in Biology and Medicine. 1995;**25**: 373-382. DOI: 10.1016/0010-4825(95)00014-U

[7] Blanco S, Figliola A, Quian QR, Rosso OA, Serrano E. Time frequency analysis of electroencephalogram series (III):Wavelet packets and information cost function. Physical Review E. 1998; **57**:932-940. DOI: 10.1103/PhysRevE.57.932

[8] Gaouda AM, Salma MMA, Sultan MR, Chikani AY. Power quality detection and classification using wavelet multiresolution signal decomposition. IEEE Transactions on Power Delivery. 1999;**14**:1469-1476. DOI: 10.1109/61.796242

[9] Rosso OA, Figliola A. Order/disorder in brain electrical activity. Revv. Mex. Fis. 2004;**50**:149-155

[10] Rosso OA, Blanco S, Yordanova J, Kolev V, Figliola A, Schurmann M, et al. Wavelet entropy: A new tool for analysis of short duration brain electrical signals. Journal of Neuroscience methods. 2001;**105**:65-75. DOI: 10.1016/S0165-0270(00)00356-3

[11] Zhao H, Wang X, Wang H. Feature selection using Relative Wavelet Energy for Brain –Computer Interface design. In: Proceedings of 2nd International Conference on Bioinformatics and Biomedical Engineering (ICBBE 2008); 16–18 May 2008. p. 1434–1437. DOI: 10.1109/ICBBE.2008.687.

[12] Subha DP, Paul KJ. Classification of EEG signals in normal and depression conditions by ANN using RWE and signal entropy. Journal of Mechanics in Medicine and Biology. 2012;**12**:1240019. DOI: 10.1142/S0219519412400192

[13] Muralidhar GB, Niranjan UC, Subha DP. Automated classification of depression EEG signals using wavelet entropies and energies. Journal of Mechanics in Medicine and Biology. 2016;**16**:1650035. DOI: 10.1142/S0219519416500354

[14] Pincus SM, Gladstone IM, Ehrenkranz RA. A regularity statistic for medical data analysis. Journal of Clinical Monitoring. 1991;**7**:335-345. DOI: 10.1007/BF01619355

[15] Higuchi T. Approach to an irregular time series on the basis of the fractal theory. Physica D: Nonlinear Phenomena. 1988;**31**:277-283. DOI: 10.1016/0167-2789(88)90081-4

[16] Acharya UR, Sudarshan VK, Adeli H, Santhosh J, Koh JEW, Subha DP, et al. A novel depression diagnosis index using nonlinear features in EEG signals. European Neurology. 2015;**74**:79-83. DOI: 10.1159/000438457

[17] Muralidhar BG, Shreya B, Lim WJE, Niranjan UC, Subha DP, Paul KJ. Automated classification of depression EEG signals using DCT and nonlinear dynamics. Journal of Medical Imaging and Health Informatics. 2015;**5**:635-640. DOI: 10.1166/jmihi.2015.1418

[18] Alan W, Jack BS, Harry LS, John AV. Determining Lyapunov exponents from a time series. Physica D: Nonlinear Phenomena. 1985;**16**:285-317. DOI: 10.1016/0167-2789(85)90011-9

[19] Theiler J, Eubank S, Longtin A, Galdrikian B, Farmer JD. Testing for nonlinearity in time series: The method of surrogate data. Physica D: Nonlinear Phenomena. 1992;**58**:77-94. DOI: 10.1016/0167-2789(92)90102-S

Chapter 6

Necessity of Quantitative EEG for Daily Clinical Practice

Jesús Pastor, Lorena Vega-Zelaya and Elena Martín Abad

Abstract

The two main problems in the daily clinical practice of EEG are i) its under-use dedicated mainly to epilepsy and ii) subjectivity in *de visu* analysis. However, both problems can be overcome by using numerical tools in clinical practice that broaden the scope and introduce real objectivity to bioelectrical measurements. We have developed a method for quantitative EEG (qEEG) for daily use based on the homeostatic foundation of EEG. This method is robust, easy, and not time consuming and is arranged in two branches: the analysis of the spectral composition in each channel and synchronization. Notably, channels are arranged in differential mode. Since 2016, we have used this method for more than 4100 EEGs from scalp recordings in outpatients, epilepsy evaluation, and evaluation and monitoring in the intensive care unit (ICU). We have been able to identify numerical properties that are not visually evident in several pathologies, including COVID-19 in patients suffering encephalopathy, and have performed diagnosis in ICU patients and differentiation between epileptic and non-epileptic spells or minimum cognitive states. The use of numerical variables across successive recordings in the same patient has proven to be of great utility. We propose that qEEG use should be expanded globally for daily clinical practice.

Keywords: encephalopathy, epilepsy, fast Fourier transform, numerical methods, psychogenic non-epileptic seizures, spectral entropy, synchronization

1. Introduction

Electroencephalography (EEG) is one of the oldest diagnostic methods currently used in medicine. It was described one century ago by the German psychiatrist Hans Berger [1]. Since then, its use has rapidly spread, and practically every hospital in the world has an EEG device. However, although EEG is probably the fastest, cheapest, and most straightforward method to obtain neurophysiological information from the human brain in a non-invasive way [2], its use is sometimes excessively restricted to the diagnosis of epilepsy, even in patients in whom the level of consciousness should be carefully evaluated [3]. Nevertheless, it should be always remembered that the primary function of the cerebral cortex is to exchange information by generating bioelectrical signals, not only in epilepsy.

In addition to this excessive restriction to epilepsy, EEG is sometimes reported as a subjective method, depending strongly on the interpreter [4–8]. There has been an attempt to reduce inter-rater variability among interpreters by the introduction of a consensus for EEG interpretation [9–13]. Although these consensus and classifications introduce some objectivity in EEG analysis, the variables are mostly qualitative or binary (present/absent), and specificity is therefore not very high.

In the past few decades, clinicians, neuroscientists, mathematicians, physicists, and engineers (among other experts) have sought a way to overcome these flaws to increase the true field of EEG application and reduce subjectivity in diagnosis. Some research has been devoted to combining EEGs and imaging techniques, mainly magnetic resonance imaging [14–16]. Another direction has been to develop mathematical tools to increase the reliability and deepen the information extracted from neurophysiological recordings. Collectively, this approach is called quantitative EEG (qEEG [17]). A canonical formulation toolbox is not needed to define qEEG. Instead, we can say that every EEG recording that uses any kind of post hoc numerical method to obtain a result about frequency spectrum, synchronization, network dynamics, or anything else of bioelectrical magnitude can be considered qEEG. This approach has been increasing in popularity since the introduction of digitalization in electroencephalography. As an example, in the past few decades, the number of papers referenced in PubMed with the word "qEEG" in the title/abstract rose from two in 1979, to 31 in 1999, and 76 in 2019 (a factor of x38!). However, the use of qEEG in clinical practice is far from being generalized, with the exception of ICUs where long-term monitoring (continuous EEG, cEEG) and qEEG are slowly increasing [18–25].

The time required for a cEEG review is one of the most commonly given reasons for the use of qEEG in the ICU and other diagnostic fields. However, we have taken a different approach to qEEG during cEEG or standard EEG for ambulatory/hospitalized patients: instead of just simplifying seizure detection (another manifestation of the excessive focus on epilepsy), our aim is to obtain a comprehensive and efficient view of the bioelectrical brain physiology/physiopathology in the most objective way. To do this, we have developed qEEG using classical mathematical methods, but in a neurophysiologically and clinically oriented fashion [26].

In this chapter, we want to describe in detail the physiological basis of qEEGs, the method implemented for its quantification, and provide some examples of its use.

2. Physiological basis of qEEG

We have adopted the assumption that EEG is founded in a homeostatic system to obtain the main variables of our method [26–28]. This aspect is of extraordinary relevance because its specific application in different pathologies will lead to specific changes in the different numerical variables obtained. EEG consists of the multivariate spatio-temporal determination of the electrical potentials generated by the brain and recorded on the surface of the scalp. The oscillatory activity of the EEG, in clinical practice, is divided into four bands, depending on its oscillation frequency: delta (δ; 0–4 Hz), theta (θ; 4–8 Hz), alpha (α; 8–13 Hz), and beta (β; 13–30 Hz). Frequencies above 30 Hz, although very important in cognitive research, are not customarily used in clinical practice.

Briefly, the regulation of different bands is given by the following systems [29]:

- Delta. The hyperpolarization of the thalamic-cortical (TC) interstitial pacemaker cells by the nucleus reticularis, together with a lower excitatory effect of the ascending activating reticular system (AARS), releases the spontaneous activity of cortical cells, oscillating at <4 Hz. This oscillation throughout the putamen/globus pallidus inhibits the brainstem nuclei responsible for AARS.

- Theta. There are two theta-generating systems: i) activation of the nucleus reticularis (activated by the AARS) and inhibiting TC neurons and ii) through the mesolimbic system, which includes multiple afferents from the entorhinal cortex, hippocampus, amygdala, septum and anterior cingulate cortex; activating the intralaminar thalamic nuclei, and projecting to layer I of the cerebral cortex.

- Alpha. TC neurons spontaneously oscillate at 6–12 Hz and regulate the excitation of large cortical areas through thalamic-cortico-thalamic re-entry loops. The membrane potential of these cells determines the frequency of oscillation and is regulated by synaptic inputs from the AARS (which includes the intralaminar nuclei of the thalamus), mainly from the brainstem and cortical inputs. Within the cortex, this activity is propagated from some nodes of special importance through interneuronal connections.

- Beta. Originates primarily from cortico-cortical interactions and is facilitated by diffuse activation of the AARS and depolarization of TC cells, allowing the free transfer of information from the sensory systems through the thalamus to the cerebral cortex.

This complex neuroanatomical homeostatic system is probably genetically determined and regulates basal levels of local synchronization, global interactions between different regions, spectral composition, and periodic signal space sampling [30–33].

One of the main limits in our approach is to maintain a close relationship between numerical magnitude variation and the underlying anatomo-functional system. For example, an increase in cortical activity (e.g., a seizure) must always be associated with an increase in β and probably α bands. Obviously, it does not preclude an increase in slower bands, but rising activity in faster bands is mandatory [26].

3. Quantified EEG

There are two branches of analysis: power spectra and synchronization. For both of them, dynamic (i.e., varying along the time) and mean measurements (i.e., mean spectra or mean graph of synchronization) are obtained. The process is summarized in **Figure 1**.

The process used for qEEG followed these steps:

Different length raw records are exported from the EEG device (EEG32, NeuroWorks, XLTEK®, Oakville, ON, Canada) to an ASCII file. Usually, artifacts are excluded by the export of several artifact-free chunks, which are later combined for analysis. We have shown that this process does not changes the main properties analyzed (see below). This process (exportation to an ASCII file) would probably be different for other EEG suppliers, but we have not assessed this possibility. We have computed the export time (t_{export}) as a function of the ASCII file size (S) and obtained a linear expression by means of least-square fitting (r = 0.9947):

$$t_{export}(s) = 0.43S(MB) + 4.13 \quad (1)$$

Although the raw recordings were digitized at 512 or 1024 Hz, we down-sampled to 128 or 256 Hz.

Exported files are digitally are filtered by a sixth-order Butterworth digital filter between 0.5 and 30 Hz.

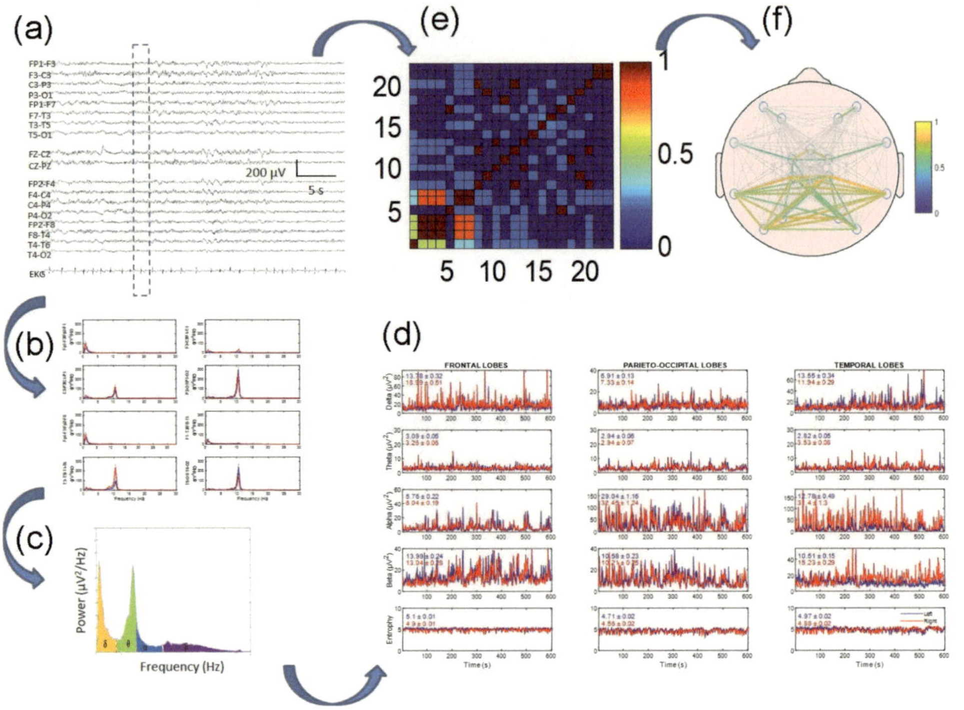

Figure 1.
Method of electroencephalogram (EEG) quantification in two branches: Power spectra (b–d) and synchronization (e,f). (a) Raw EEG tracing. The discontinuous rectangle shows the moving window used for analysis; (b) power spectra for each channel; (c) areas for delta, theta, alpha, and beta bands under the spectrum are highlighted in different colors; (d) dynamics of the four bands (and entropy in the lower row) for every lobe. Mean and SEM values for each tracing are displayed inside each graph. Red and blue lines indicate right and left hemispheres, respectively; (e) correlation matrix for the window; (f) mean correlation computed for all recordings [34].

A differential EEG montage is then reconstructed. Topographic placement of channels is defined on the scalp as the midpoint between the electrode pairs defining the channel; e.g., the Fp1–F3 channel would be placed at the midpoint of the geodesic between the Fp1 and F3 electrodes.

All recording can be divided into different lengths of moving windows (1–5 s each) with different overlaps (between 0 to 50%, but usually 10%). The total length used during the fast Fourier transform (FFT) is directly related to frequency precision in the power spectrum (PS). Overlap is used to minimize the border effect produced by windowing [35].

For each window (n) and frequency (k), we computed the fast Fourier transform (FFT) of the voltage ($V^m(n)$) obtained from each channel (m) to obtain the power spectrum ($S^m_{n,k}$, in $\mu V^2/Hz$). We used the expression:

$$S^m_{n,k} = \sum_{n=0}^{N-1} V^m(n) e^{-i\frac{2\pi}{N}kn}; m = Fp_1, F_3, \ldots \quad (2)$$

We also computed Shannon's spectral entropy (SSE) according to:

$$SSe^m_k = -\sum_{k=0}^{F} p_k \log_2 p_k \quad (3)$$

where F is the maximum frequency computed and p_k is the probability density of S, obtained from the expression:

$$p_k = \frac{S^m_{n,k}}{\sum_{k=0}^{F} S^m_{n,k} \Delta k} \tag{4}$$

We computed the area under the $S^m_{n,k}$ according to the classical segmentation of EEG bands. We used the expression:

$$A_j(k) = \sum_{k=inf}^{sup} S^m_n(k)\Delta k; j = \delta, \theta, \alpha, \beta \tag{5}$$

The expression *sup* refers to the upper limit of each EEG band.

The absolute value of Pearson's correlation coefficient (ρ) is computed for each pair of channels (i,j) according to the expression:

$$\rho^k_{ij} = \frac{\sum_{k=1}^{N_{window}} (x_i(k) - \overline{x}_i) \sum_{k=1}^{N_{window}} (x_j(k) - \overline{x}_j)}{\sqrt{\sum_{k=1}^{N_{window}} (x_i(k) - \overline{x}_i)^2 \sum_{k=1}^{N_{window}} (x_j(k) - \overline{x}_j)^2}} \tag{6}$$

where N_{window} is the number of points included in a window (usually 128) and $\overline{x}_i, \overline{x}_j$ represents the mean of both channels.

The mean value of all windows is then computed, obtaining the mean correlation matrix.

Areas of the same band are grouped by cerebral lobes. In the case of the left hemisphere (shown as an example), we grouped the frontal $F = \left\{\frac{(Fp_1-F_3)+(F_3-C_3)+(Fp_1-F_7)}{3}\right\}$, parieto-occipital $PO = \left\{\frac{(C_3-P_3)+(P_3-O_1)+(T_5-O_1)}{3}\right\}$, and temporal $T = \left\{\frac{(Fp_1-F_7)+(F_7-T_3)+(T_3-T_5)+(T_5-O_1)}{4}\right\}$. Channels from the right hemisphere were grouped accordingly. These areas, for both bands (j) and lobes (r), $A^r_j(t); r = F, PO, T$, are plotted as time functions and compared between the hemispheres. The same groups were used to compute *SSe*.

The total time of analysis ($t_{analysis}$) is obtained from this linear expression, which was obtained from least-square fitting:

$$t_{analysis}(s) = 0.32S(MB) + 46.3 \tag{7}$$

From expressions 1 and 7, for a typical 88 MB file (10 min record), we can estimate the time spent in export + analysis as less than 2 min.

We can optionally introduce two time-markers to define different states (e.g., pre-ictal, ictal, and post-ictal periods) in order to statistically compare the changes.

We can optionally export the numerical results to an Excel® file (e.g., mean ± SEM of power, synchronization; and SSe for channels, lobes, and hemispheres). This last step is the most time-consuming (up to 3 min for a custom-length file).

Numerical analysis of EEG recordings was performed with custom-made MATLAB® software (MathWorks, Natick, MA, USA).

For power spectra as well as for synchronization, we can represent measurements either as dynamic time-dependent variables (**Figure 1d**), or as the mean values, averaged over the file (**Figure 1b,f**). Therefore, although complementary, information obtained from both kinds of computations must be interpreted differently. In other words, average measurements are only useful if the stationarity of the record is evident (e.g., basal recordings, well-defined phases of sleep, etc.).

4. Robustness of the method

A very important aspect of any numerical method is its robustness, e.g., the evaluation of the method wherein the results obtained are found to be reliable, even when performed under slightly varied conditions. It is the ability of a method to remain unaffected when slight variations are applied. It is extremely important to check trials of the numerical method within the same group of EEG records under the different conditions of i) down sampling, ii) windowing, or iii) overlapping. Moreover, it is important to check whether synchronization measures are affected by the global analysis of different, non-consecutive chunks.

For this purpose, we selected EEG recordings of five minutes in length, from six control individuals (without any neurological or psychiatric pathology, between 20 and 30 years old, and with no pharmacological treatment). We analyzed each EEG under different conditions, namely:

- Down-sampling at 128 and 512 Hz, (f is for frequency).
- Windows of 1, 3 or 5 s (w).
- Overlapping at 10, 20, or 40% (o).

Overall, we had 18 combinations for frequency/windows/overlapping (f,w,o).

The structure of EEG for each patient (p_i, i = 1,2, ... 6) can be described by a 10-element vector as:

$$p_i = (\delta_i^l, \theta_i^l, \alpha_i^l, \beta_i^l, \rho_i^l, \delta_i^r, \theta_i^r, \alpha_i^r, \beta_i^r, \rho_i^r); l = left, r = right \tag{8}$$

Obviously, every structure can be described as $p_i(f, w, o)$. A robust method should not affect the structure of EEG for the same patient, irrespective of changes in absolute values. For each patient, we have plotted along the x-axis (coordinates of EEG structure) the normalized band values and correlations for all of the 18 combinations (**Figure 2**).

From **Figure 2**, we can observe that different combinations cannot affect the structure of EEG.

The effect of multiple compositions of the analyzed file on synchronicity was assessed as follows: an EEG record of 3 min was analyzed. Then, the same record was exported in three different chunks, and a new analysis was performed on a recombined file with the parts randomly ordered (1,3,2/2,1,3/2,3,1/3,1,2 or 3,2,1). We did not observed any difference in synchronization between the whole record and a recombined one (not shown).

In summary, these results demonstrate that the method is highly robust, at least for the limits addressed.

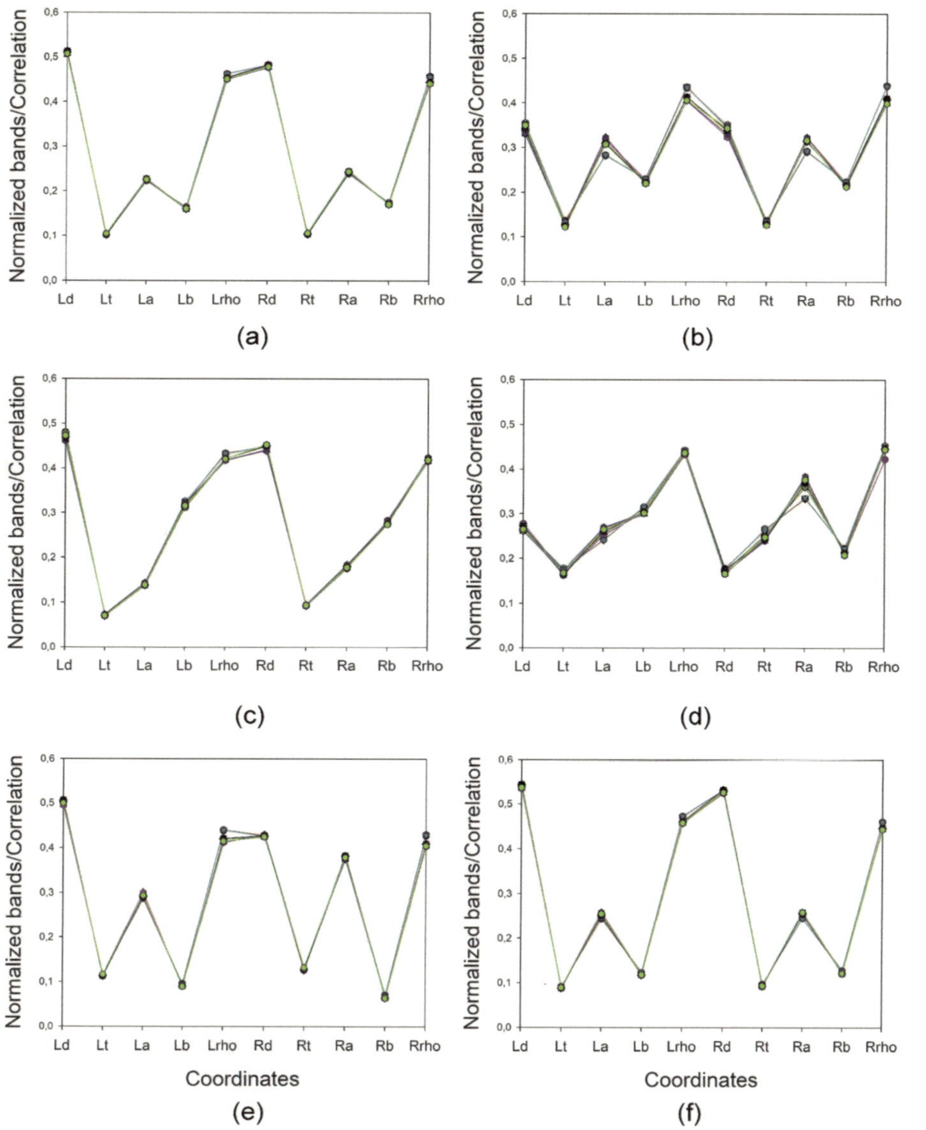

Figure 2.
Structures of EEG for the 18 combinations of variables for (a) patient #1, (b) patient #2, (c) patient #3, (d) patient #4, (e) patient #5 and (f) patient #6. L = left, R = right, d = delta, t = theta, a = alpha, b = beta, and rho = correlation.

5. Some examples of the utility of qEEG

It is out of the scope of this chapter to illustrate the objective and quantified difference between the classical, *de visu*, analysis of EEG and qEEG. Instead, we will provide some examples of qEEG application to real clinical problems and the results. In these examples it will be shown that information obtained by qEEG clearly exceeds the possibility of *de visu* analysis.

5.1 Differentiation between periodic patterns and seizures

Patients in the ICU usually undergo a limited clinical neurological examination because of either structurally or functionally altered conditions of the central nervous system (CNS) or due to the effects of drugs used for sedoanalgesia [36]. Therefore, we can evaluate brain function in these conditions by EEG. However, a dynamic evolution of injury is frequently observed in critically ill patients due to the occurrence of epileptic seizures (ES), status epilepticus (SE), or other brain insults [37, 38]. In this sense, cEEG allows the functional assessment of the cerebral cortex in real time for prolonged periods. It has been proven to be an extraordinarily useful tool for detecting electrographic seizures and non-convulsive epileptic status (NCES), modifying treatment and assessing the functional prognosis [10, 11, 39–42].

The particularity of cEEG records of patients in the ICU comprises a high frequency of artifacts and frequently observed rhythmic and periodic patterns, which are easily confused with seizures. Both can be difficult to interpret not only in a raw EEG but also with the qEEG tools currently used [43, 44]. As indicated above, the use of multiple drugs acting on the CNS and primary and secondary injuries profoundly affect the bioelectrical brain dynamics. Therefore, it can be quite difficult to differentiate between bursts of periodic activity (BPA) and true ES/SE. The main problem is that EEG patterns analyzed *de visu* do not always exhibit the sharp morphology of ES/SE of not-ICU patients.

According to the ILAE, an ES is a transient occurrence of signs and/or symptoms due to abnormal excessive or synchronous neuronal activity in the cortex of the brain [45]. Signs/symptoms are usually excluded in critically ill patients, but the excessive activity of the cortex is mandatory for a positive identification. We have used this pathophysiological feature for the numerical definition of ES. Therefore, we can use this method to exclude epilepsy in those cases where α/β activity does not change (or even decreases) during the event (see below). The limits of change for the different bands (**Table 1**) can be used to distinguish PBA from ES.

From this table we can observe that the superposition is very high for δ bands (i.e., this band is not discriminative), low for θ and *Se,* and practically null for α and β bands. Therefore, the intervals for increments of normalized activity defining an ES in these types of patients are as follows (excluding superposition and rounding): $\delta_F = [119, 166]$; $\theta_F = [173, 264]$; $\theta_{PO} = [168, 248]$; $\theta_T = [151, 274]$; $\alpha_F = [159, 244]$; $\alpha_{PO} = [159, 244]$; $\alpha_T = [159, 244]$; $\beta_F = [141, 374]$; $\beta_{PO} = [146, 262]$; $\beta_T = [141, 374]$; $Se_F = [97, 110]$; $Se_{PO} = [98, 107]$; $Se_T = [98, 104]$.

We have defined the numerical features of ES and BPA in critically ill patients using the pathophysiological definition of epilepsy. This will facilitate its identification in clinical practice, allowing a precocious and more adequate treatment.

5.2 Specific characterization of encephalopathy in SARS-CoV-2 patients

Neurological complications in COVID-19-infected patients have been reported. The CNS effects reported include encephalitis, toxic encephalopathy, ageusia and anosmia, headaches, or acute cerebrovascular disease [46–52]. The mechanisms of CNS infection in the pathophysiology of COVID-19 are still debated, and it has been proposed to result from direct invasion through the blood–brain barrier, a neuronal pathway, hypoxia damage, immune-response mediated injury, or angiotensin-converter enzyme 2 activity, among other possibilities [47, 53, 54]. Encephalopathy refers clinically to state of impaired cognition, generally acute or subacute [55]. Descriptions *de visu* of EEG are based on classical analysis by visual inspection and not on specific features [56–58].

We applied our method of qEEG to patients discharged from the ICU after COVID-19 infection [34]. We used two control groups from patients previously studied in our hospital: (i) patients with infectious toxic encephalopathy (ENC) and (ii) patients after cardiorespiratory arrest (CRA), as an example of severe hypoxic insult to the CNS.

Variable	State	Frontal	Sup (%)	Parieto-occipital	Sup (%)	Temporal	Sup (%)
Delta	ES	111.8–234.0	55.4	123.8–276.8	91.3	118.4–247.6	86.1
	BPA	166.3–370.4		137.1–278.7		136.3–253.9	
Theta	ES	158.2–264.0	14.2	166.7–247.7	1.6	139.2–230.5	19.9
	BPA	92.8–173.2		84.3–168.0		93.1–157.4	
Alpha	ES	158.8–244.0	0	146.0–248.6	0	144.3–243.5	0
	BPA	75.0–137.8		79.7–134.1		82.7–135.6	
Beta	ES	141.9–373.6	0	146.7–261.8	0	136.5–274.4	10.4
	BPA	77.1–137.8		82.2–137-2		95.1–150.8	
Entropy	ES	95.5–109.5	10.7	96.2–106.7	12.4	96.1–104.8	19.5
	BPA	78.9–97.0		84.3–97.5		82.3–97.8	

Table 1.
Inter-percentile 25–75 intervals for bands and lobes in BPA and ES. Superposition is indicated with respect to the ES interval [26].

Visually, in COVID-19, EEG records the apparent absence of delta/theta activity conferred a near-physiological aspect to the recordings. Despite the different visual aspects appearing between ENC and COVID-19, the mean spectra by channel were quite similar. Examples of typical recordings are shown in **Figure 3**.

We assessed the specific differences for each band in ENC, COVID-19, and CRA patients. Subsequently, we performed one-way ANOVA (ANOVA on ranks when normality failed) for each lobe and band, and for SSe and synchronization (**Figure 4**).

The pattern of ENC and CRA was clearly different for all bands. However, the COVID-19 group was not completely different from the ENC and CRA groups, although it was evidently observed that the distribution is between the two extreme groups. Nonetheless, the behavior of temporal lobes clearly differs for the ENC and COVID-19 groups for the δ, α, and β bands. Contrary to EEG bands, SSe was higher for the COVID-19 group and lower for the ENC group. In fact, SSe was different for all groups at all lobes. This result may be surprising considering the kind of spectra shown in **Figure 3c**, where the distribution was apparently more complex. However, the presence of α and β bands (scarcely present in the CRA group) increased the SSe for COVID-19 patients.

Finally, although ρ was not as different between groups as SSe, a clear difference was seen in the hemispheric synchronization and frontal lobes of ENC and COVID-19 patients, with lower synchronization for the latter group.

In summary, we have demonstrated that qEEG can differentiate between encephalopathy types, and we have described the numerical features of each. In this context, we show that COVID-19 patients display EEG structures that are truly distinguishable from those of both infectious toxic encephalopathy and encephalopathies of patients who experience severe hypoxic conditions. Significantly, the EEG pattern of COVID-19 patients was between those of the ENC and CRA groups. Therefore, it can be speculated that hypoxia may show some participation in this electroclinical entity. However, the EEG structures of the CRA and COVID-19 groups were different enough to consider that other factors besides hypoxia must be responsible for the bioelectrical pattern.

It is extremely relevant to bear in mind that COVID-19 patients showed mild to severe cognitive symptoms despite *de visu* quasi-normal recordings. However, the severe numerical alterations of temporal lobes spectra, structure of SSe and

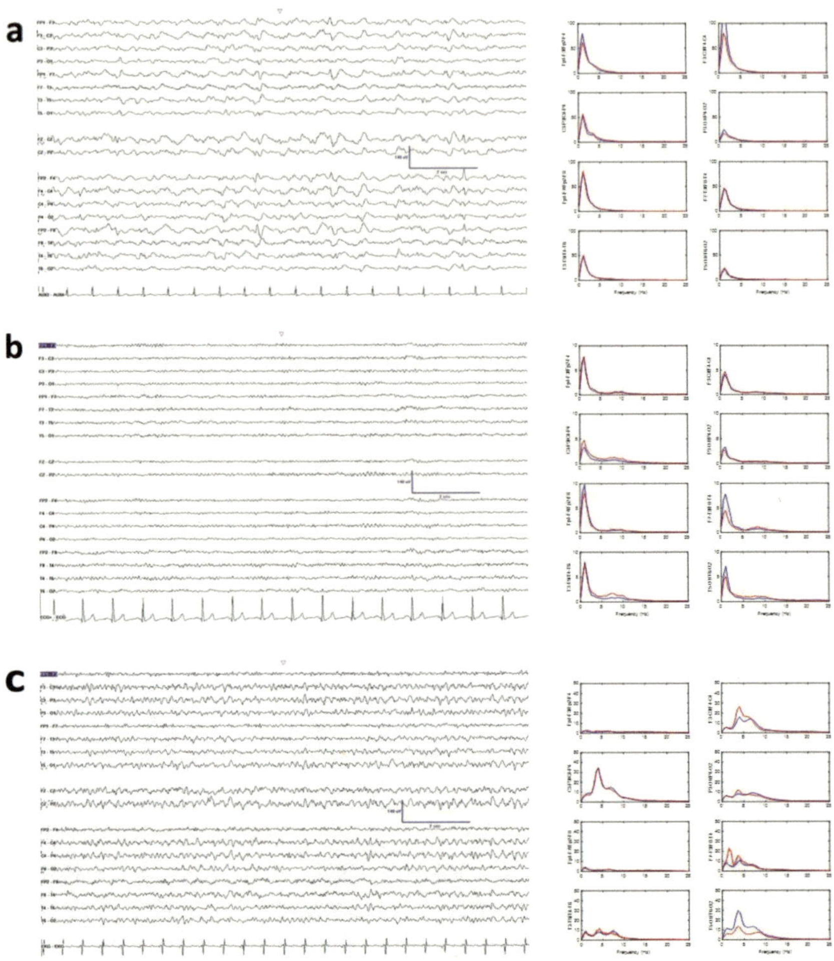

Figure 3.
Examples of raw recordings: (a) encephalopathy (ENC), (b) COVID-19, and (c) cardiorespiratory arrest (CRA). Right column shows mean spectra for channels. Red and blue lines indicate right and left hemispheres, respectively. Y-axis units in $\mu V^2/Hz$ [34].

synchronization were highly different and are probably explained by symptomatology with quasi-normal EEG traces.

5.3 Continuous EEG monitoring in ICU

Long-term EEG monitoring in the ICU (cEEG) has been one of the main developing fields of electroencephalography in recent years. There are several methods used for EEG monitoring, and most of them share a similar philosophy: to identify the presence of seizure/status epilepticus to make the information of monitoring easy and fast. As stated above, we think that this reductionist approach impoverishes the scope of electroencephalography.

We systematically used qEEG during cEEG for the below indications:

- Monitoring the presence of ES/SE. More useful than identifying the presence of ES, qEEG is essential to discriminate between epileptic and non-epileptic patters, as we stated above.

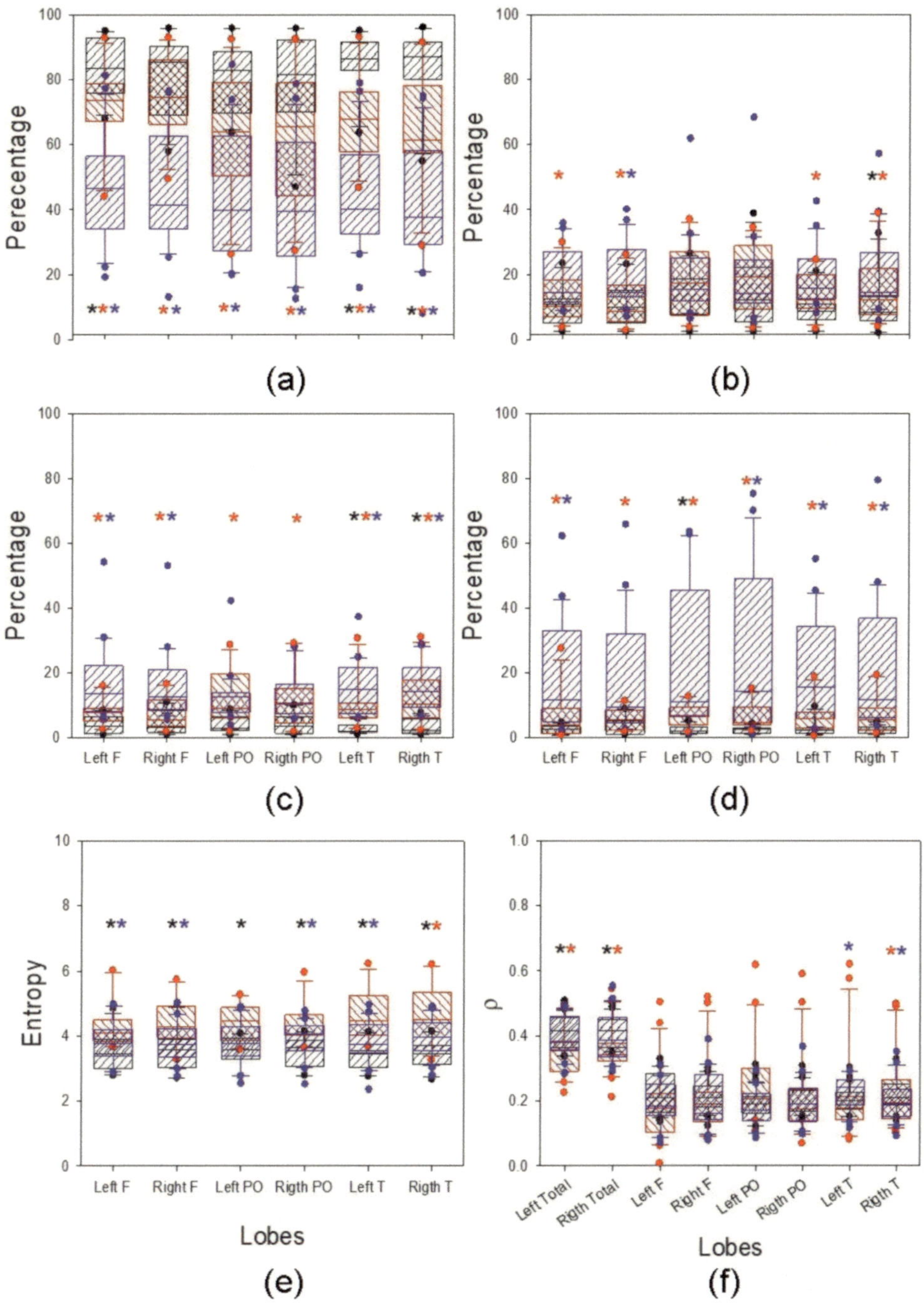

Figure 4.
Box plots showing the comparison of EEG structures for different bands: (a) delta, (b) theta, (c) alpha, and (d) beta, (e) Se; (f) ρ. striped black box: ECN; striped red box: COVID-19; striped blue box: CRA; black asterisk: Difference between ENC and COVID-19; red asterisk: Difference between ENC and CRA; blue asterisk: Difference between COVID-19 and CRA.

- Titration of sedation/anti-epileptic drugs (AEDs). Identifying changes in background from the variations in spectra or mean values of specific EEG bands is mandatory to increase/decrease the dose of sedation. Additionally, AEDs are adjusted with the help of qEEG, although *de visu* inspection of irritative activity is mandatory.

- Long-term recordings in patients with alterations of consciousness without sedo-analgesia to evaluate brain physiology. Severe encephalopathy can induce a low-consciousness level. These recordings typically lack irritative activity, and changes in the background are slow. However, these changes correlate with the level of consciousness and predict the outcome. Therefore, it is important identify changes to adjust treatment as soon as possible, especially to avoid unnecessary therapeutic actions.

In this way, we have performed more than 250 cEEG + qEEG in the ICU in the last five years. This is a time-consuming task (especially considering that every cEEG takes 1.4–4.3 days (inter-percentile 25–75 range)), but the clinical utiliy is clear because the demand increased from 1.0 $\pm$ 0.2 cEEG/month in 2015 to 5.5 $\pm$ 0.8 cEEG/month in 2019.

5.4 Utility in dementia

There are numerous articles in the literature showing that the initial phases of dementia can be detected by qEEG [59–63]. We have used our numerical method in patients with either minimum cognitive impairment (MCI) or aphasia. Obviously, the *de visu* analysis of raw recordings shows only a nearly normal aspect or low-voltage. However, numerical analysis can show very relevant facts that are not observable by eye (**Figure 5**).

At this time, we are conducting a study to identify specific properties of different pathologies (sub-types of primary aphasia, Alzheimer disease, vascular dementia, etc). Although we have not yet defined different groups of features specific to each pathology, what is clear from the above figure is that connectivity is a magnitude that is affected early and consistently in most cognitive alterations.

5.5 Other examples of qEEG utility

Finally, we provide two more representative examples of diagnosis highly aided by the use of qEEG.

Case 1. A 17-year-old male patient with severe cognitive and behavioral impairment, secondary to severe epileptic encephalopathy due to refractory epilepsy in childhood after a central nervous system infection. Daily seizure frequency, with countless seizures per day. In treatment with zonisamide (400 mg), valproic acid (1200 mg), oxcarbazepine (1400 mg), and clobazam (10 mg) daily. Rectal diazepam 10 mg if required. A video of EEG is performed in which it is observed that during sleep, the patient exhibits several episodes of lateral head movement and growls. *De visu* recordings (**Figure 6a**) can be described as global desynchronization. However, the dynamics of the EEG bands show a decrease (practically total) for all the bands (**Figure 6b**). As stated above, epilepsy is expected to be accompanied by an increase in cortical activity (α and β bands). Therefore, this event cannot be identified as epileptic.

Although the patient also presented true epileptic events, the number of these was substantially lowered and AED was adjusted according to the real number of ES. The final diagnosis for this event was a nonepileptic behavioral disorder, secondary to severe epileptic encephalopathy.

Case 2. A 22-year-old female diagnosed with epilepsy at the age of 14 years and with anxious-depressive illness from the age of 20 years. The applied treatment was lamotrigine (100 mg/day) and clonazepam (0.5 mg/8 h). Seizures occurred every 2–3 days, described as the perception of black dots in the visual field, weakness, loss of consciousness and muscle tone, loss of balance and falling to the ground. During

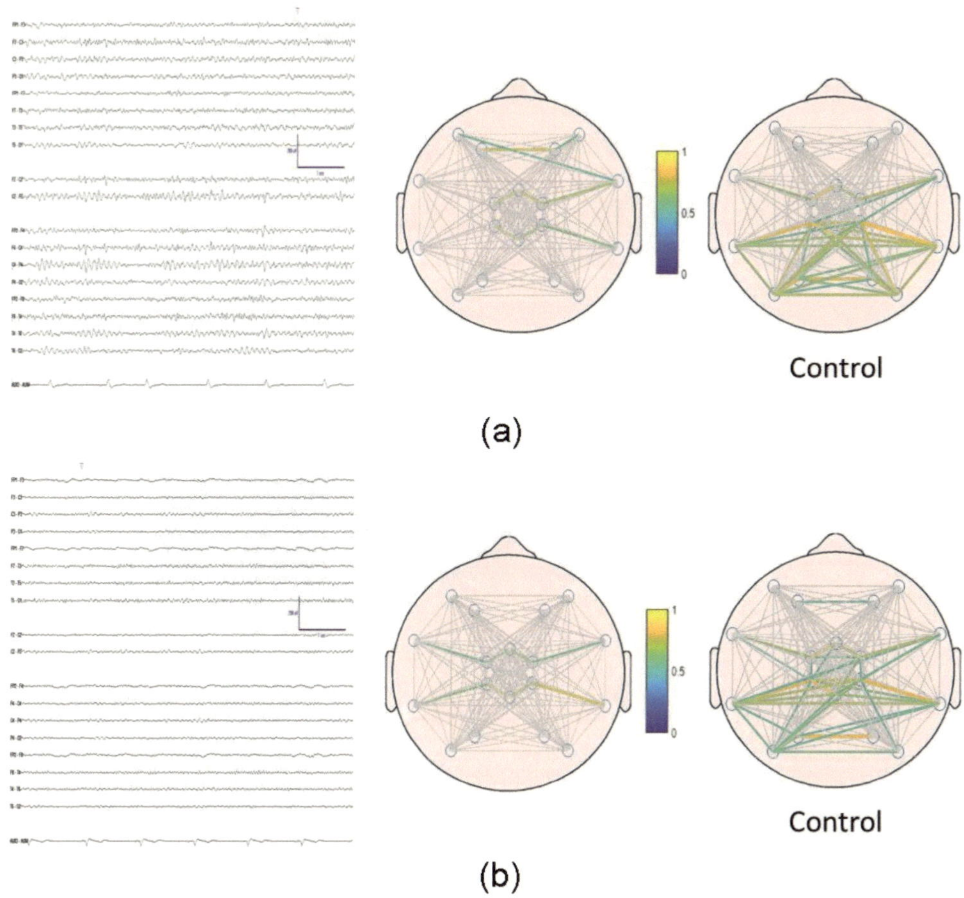

Figure 5.
Initial steps of dementia. (a) Example of a male with primary aphasia and (b) a female with MCI. Left column = raw recordings of both patients; middle column = connectogram for patients; right column = connectogram of a control volunteer of the same sex and age (± 5 years).

telemetry, we recorded one episode of black dot perception followed by a loss of consciousness while in bed. *De visu* EEG recordings showed no significant changes (**Figure 7a**). However, the dynamic variation of EEG bands indicated a generalized decrease, except for the occipital α band, which increased after the eyes closed (**Figure 7b**).

The final diagnosis was psychogenic non-epileptic seizure (PNES), and the AEDs were slowly removed.

6. Discussion

Numerous methods have been used for qEEG, although they are rarely used in daily clinical practice. There is, therefore, a huge gap between the promising (even spectacular) results obtained with qEEG and its practical usefulness. To the best of the knowledge of the authors, this issue has not been systematically addressed, although it has been said that electroencephalographers have poor trust in mathematical models [2, 17].

The degree of mathematical complexity and abstraction is quite different among the methods proposed. Not all of the mathematical models can be included in the same category, and it is of extreme importance that mathematical solutions be

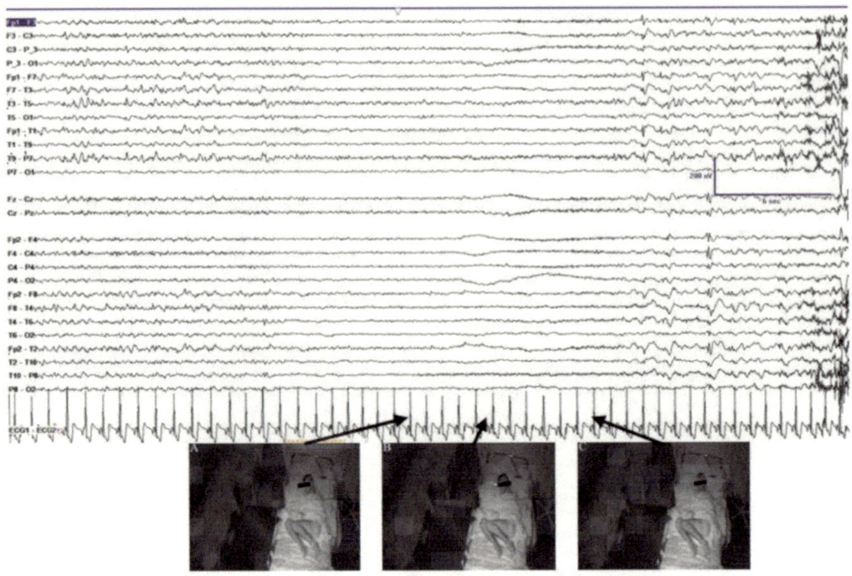

(a)

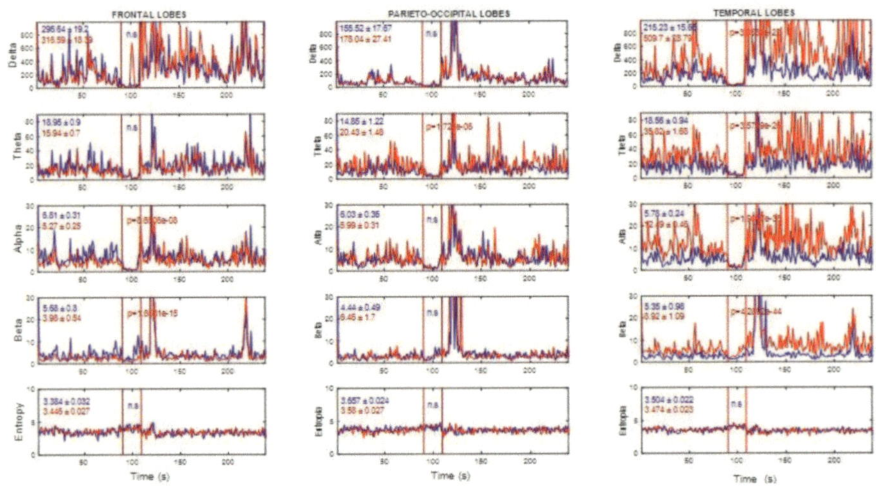

(b)

Figure 6.
Box plots showing a comparison of EEG structure for different bands: (a) raw recording during a complete event during sleep. The images under the recordings correspond to the periods indicated by arrows. (b) Lobar dynamics of EEG bands during the entire event (vertical red lines). Blue = left hemisphere; red = right hemisphere.

robust with respect to physiological assumptions [2]. Then, some procedures require relatively simple and involve straightforward methods as FFT [26, 34, 64–66] and synchronicity measurements [67, 68], approaches that clinicians are familiar with. In contrast, other numerical methods are more complex, or the mathematical approach deviates from physiological assumptions [69–74], and this probably takes neurophysiologists out of their comfort zone.

We have developed a robust method that is physiologically founded and easy to use in daily clinical practice. The tools selected are neither unique nor are they necessarily the best. Other tools (e.g., coherence) can be implemented. A careful comparison between these methods will decide the fitted procedure for each

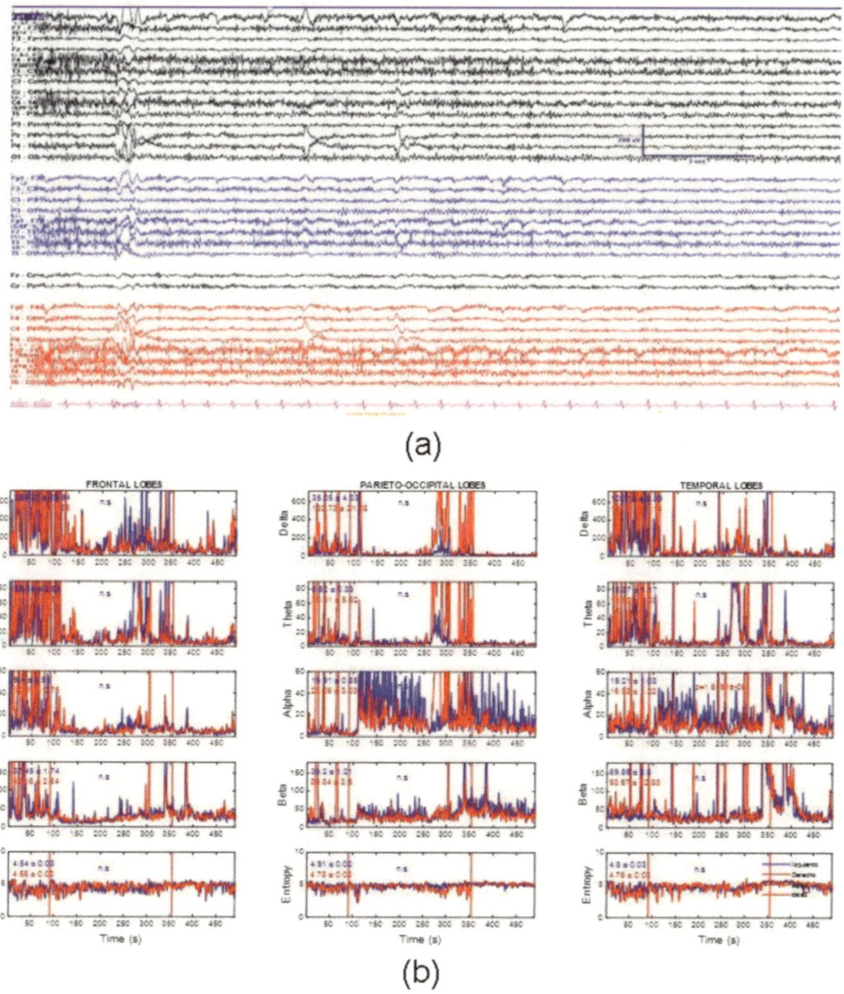

Figure 7.
Psychogenic non epileptic seizure (a) raw recordings (in transverse and double banana differential montages) during the event. (b) Lobar dynamics of EEG bands during the entire event (vertical red lines). Blue = left hemisphere; red = right hemisphere.

pathology. From the beginning of 2016, we have performed more than 4100 analyses and used this toolbox in most patients, even those with EEG apparently evident. We did not use qEEG only when the record included so many artifacts (e.g., in agitated patients) that the results would not be reliable.

The method described can be implemented to automatically differentiate between paroxysmal events and ES during the long-term monitoring of ICU patients. This feature is very relevant for clinicians because it can shorten the review time, particularly during long cEEG, and can help apply adequate therapeutic measures, avoiding pharmacological blind trials that only delay correct treatment, increasing the inefficacy of treatment and diminishing the probability of recuperation. Therefore, considering that "time is brain", a fast and accurate treatment is mandatory to increase the probability of a good outcome.

Finally, it is extremely important to keep in mind that qEEG is only a tool to help better understand and diagnose brain pathophysiology; therefore, it should not be thought that numerical analysis (at least as we use it today) is enough, without evaluation by an expert, to make an automatic diagnosis. Not all brain pathologies are likely to benefit to the same degree from analysis. For example, the method described in this chapter is not well fitted to detect low-frequency transitory waves as medium/small focal epileptiform discharges, although visual inspection can identify them very well. However, patterns that are not readily visible in *de visu* analysis (e.g., asymmetries in power spectra compositions) are easily detected.

7. Conclusions

We can summarize the conclusions of this work as follows:

- qEEG is a robust and non-time-consuming method that is able to produce numerical and objective values of several bioelectrical magnitudes with clinical significance.

- qEEG increases and facilitates diagnoses that are otherwise exceedingly difficult to obtain by *de visu* inspection. Therefore, the scope of the true applicability of EEG is expanded far beyond epilepsy.

- We propose that qEEG use should be expanded globally for daily clinical practice. Thus, clinical neurophysiologists should be informed of the methods of numerical analysis procedures.

Acknowledgements

This work was financed by a grant from the Ministerio de Sanidad FIS PI17/02193, ISCIII (Instituto de Salud Carlos III) and partially supported by FEDER (Fonds Europeen de Developpement Economique et Regional).

Conflict of interest

The authors declare no conflict of interest.

Author details

Jesús Pastor*, Lorena Vega-Zelaya and Elena Martín Abad
Clinical Neurophysiology, Hospital Universitario La Princesa, Madrid, Spain

*Address all correspondence to: jesus.pastor@salud.madrid.org

References

[1] Berger H: Uber das elektroenzephalogramm des Menschen. Arch. Psychiatr. Nervenk. 1929; 87: 527–570.

[2] Nunez PL, Srinivasan , editors. Electric fields of the brain. The neurophysics of EEG. 2nd ed. Oxford University Press; 2006.

[3] Pellinen J, Carroll E, Friedman D, Boffa M, Dugan P, Friedman DE, Gazzola D, Jongeling A, Rodriguez aj, Holmes M: Continous EEG findings in patients with COVID-19 infection admitted to a New York academic hospital system. Epilepsia; 2020; 00: 1–9; doi.org/10.1111/epi.16667

[4] Wendt SL, Welinder P, Sorensen HB, Peppard PE, Jennum P, Perona P, Mignot E, Warby SC: Inter-expert and intra-expert reliability in sleep spindle scoring. Clin Neurophysiol. 2015;126(8): 1548–56. doi: 10.1016/j.clinph.2014.10.158.

[5] Halford JJ, Shiau D, Desrochers JA, Kolls BJ, Dean BC, Waters CG, Azar NJ, Haas KF, Kutluay E, Martz GU, Sinha SR, Kern RT, Kelly KM, Sackellares JC, LaRoche SM: Inter-rater agreement on identification of electrographic seizures and periodic discharges in ICU EEG recordings. Clin Neurophysiol. 2015; 126(9):1661–9. doi: 10.1016/j.clinph.2014.11.008.

[6] Halford JJ, Arain A, Kalamangalam GP, LaRoche SM, Leonardo B, Basha M, Azar NJ, Kutluay E, Martz GU, Bethany WJ, Waters CG, Dean BC: Characteristics of EEG Interpreters Associated With Higher Interrater Agreement. J Clin Neurophysiol. 2017 Mar;34(2):168–173. doi: 10.1097/WNP.0000000000000344.

[7] Halford JJ, Westover MB, LaRoche SM, Macken MP, Kutluay E, Edwards JC, Bonilha L, Kalamangalam GP, Ding K, Hopp JL, Arain A, Dawson RA, Martz GU, Wolf BJ, Waters CG, Dean BC: Interictal Epileptiform Discharge Detection in EEG in Different Practice Settings. J Clin Neurophysiol. 2018 Sep;35(5):375–380. doi: 10.1097/WNP.0000000000000492.

[8] Jing J, Herlopian A, Karakis I, Ng M, Halford JJ, Lam A, Maus D, Chan F, Dolatshahi M, Muniz C, Chu C, Sacca V, Pathmanathan J, Ge W, Sun H, Dauwels J, Cole AJ, Hoch DB, Cash SS, Westover MB: Interrater Reliability of Experts in Identifying Interictal Epileptiform Discharges in Electroencephalograms. JAMA Neurol. 2019;77(1):49–57. doi: 10.1001/jamaneurol.2019.3531

[9] Hirsch LJ, LaRoche SM, Gaspard N, Gerard E, Svoronos A, Herman ST, Mani R, Arif H, Jette N, Minazad Y, Kerrigan JF, Vespa P, Hantus S, Claassen J, Young GB, So E, Kaplan PW, Nuwer MR, Fountain NB, Drislane FW: American Clinical Neurophysiology Society's Standardized Critical Care EEG Terminology: 2012 version. J Clin Neurophysiol. 2013 Feb;30(1):1–27. doi: 10.1097/WNP.0b013e3182784729. PMID: 23377439.

[10] Herman ST, Abend NS, Bleck TP, Chapman KE, Drislane FW, Emerson RG, Gerard EE, Hahn CD, Husain AM, Kaplan PW, LaRoche SM, Nuwer MR, Quigg M, Riviello JJ, Schmitt SE, Simmons LA, Tsuchida TN, Hirsch LJ: Critical Care Continuous EEG Task Force of the American Clinical Neurophysiology Society. Consensus statement on continuous EEG in critically ill adults and children, part I: indications. J Clin Neurophysiol. 2015 Apr;32(2):87–95. doi: 10.1097/WNP.0000000000000166.

[11] Herman ST, Abend NS, Bleck TP, Chapman KE, Drislane FW, Emerson RG, Gerard EE, Hahn CD, Husain AM, Kaplan PW, LaRoche SM,

Nuwer MR, Quigg M, Riviello JJ, Schmitt SE, Simmons LA, Tsuchida TN, Hirsch LJ: Critical Care Continuous EEG Task Force of the American Clinical Neurophysiology Society. Consensus statement on continuous EEG in critically ill adults and children, part II: personnel, technical specifications, and clinical practice. J Clin Neurophysiol. 2015 Apr;32(2):96–108. doi: 10.1097/WNP.0000000000000165.

[12] Admiraal MM, van Rootselaar AF, Horn J: International consensus on EEG reactivity testing after cardiac arrest: Towards standardization. Resuscitation. 2018 Oct;131:36–41. doi: 10.1016/j.resuscitation.2018.07.025.

[13] Claassen J, Taccone FS, Horn P, Holtkamp M, Stocchetti N, Oddo M: Neurointensive Care Section of the European Society of Intensive Care Medicine. Recommendations on the use of EEG monitoring in critically ill patients: consensus statement from the neurointensive care section of the ESICM. Intensive Care Med. 2013 Aug;39(8):1337–51. doi: 10.1007/s00134-013-2938-4.

[14] Robertson JA, Thomas AW, Prato FS, Johansson M, Nittby H: Simultaneous fMRI and EEG during the multi-source interference task. PLoS One. 2014 Dec 9;9(12):e114599. doi: 10.1371/journal.pone.0114599

[15] Klein C, Hänggi J, Luechinger R, Jäncke L: MRI with and without a high-density EEG cap—what makes the difference? Neuroimage. 2015;106:189–97. doi: 10.1016/j.neuroimage.2014.11.053.

[16] Omidvarnia A, Kowalczyk MA, Pedersen M, Jackson GD: Towards fast and reliable simultaneous EEG-fMRI analysis of epilepsy with automatic spike detection. Clin Neurophysiol. 2019 Mar;130(3):368–378. doi: 10.1016/j.clinph.2018.11.024.

[17] Vega-Zelaya L, Garnés-Camarena O, Sánz-García A, Ortega GJ, Pastor J: Mathematical foundations of quantified electroencephalography. Clinical Advances in Neurophysiology. Chapter 3. Avid Science. 2016; February, 10.

[18] Hirsch LJ: Continuous EEG monitoring in the intensive care unit: an overview. J Clin Neurophysiol. 2004; 21: 332–340.

[19] Kennedy JD, Gerard EE: Continuous EEG monitoring in the intensive care unit. Curr Neurol Neurosci Rep. 2012; 12: 419–428.

[20] Ney JP, van der Goes DN, Nuwer MR, Nelson L, Eccher MA: Continuous and routine EEG in intensive care: utilization and outcomes, United States 2005–2009. Neurology. 2013; 81: 2002–2008.

[21] Sutter R, Stevens RD, Kaplan PW: Continuous electroencephalographic monitoring in critically ill patients: indications, limitations and strategies. Crit Care Med. 2013; 41: 1124–32.

[22] Sackellares JC, Shiau DS, Halford JJ, LaRoche SM, Kelly KM: Quantitative EEG analysis for automated detection of nonconvulsive seizures in intensive care units. Epilepsy Behav. 2011, 22, S69–S73.

[23] Dericioglu N, Yetim E, Bas DF, Bilgen N, Caglar G, Arsava EM, Topcuoglu MA: Non-expert use of quantitative EEG displays for seizure identification in the adult neuro-intensive care unit. Epilepsy Res. 2015, 109, 48–56.

[24] Haider HA, Esteller R, Hahn CD, Westover MB, Halford JJ, Lee JW, Shafi MM, Gaspard N, Herman ST, Gerard EE, Hirsch LJ, Ehrenberg JA, LaRoche SM: Critical Care EEG Monitoring Research Consortium. Sensitivity of quantitative EEG for seizure identification in the intensive care unit. Neurology. 2016 Aug 30;87 (9):935–44. doi: 10.1212/

WNL.0000000000003034. Epub 2016 Jul 27. PMID: 27466474; PMCID: PMC5035158.

[25] Zubler F, Bandarabadi M, Kurmann R, Steimer A, Gast H, Schindler K: Quantitative EEG in the Intensive Care Unit. Epileptologie. 2016; 33: 166–172

[26] Vega-Zelaya L, Martín Abad E, Pastor J: Quantified EEG for the characterization of epileptic seizures versus periodic activity in critically ill patients. Brain Sci. 2020, 10, 158; doi: 10.3390/brainsci10030158.

[27] John ER, Ahn H, Prichep, LS, Trepetin M, Brown D, Kaye H: Developmental equations for the electroencephalogram. Science 1980, 210, 1255–1258.

[28] John ER, Prichep LS, Easton P: Normative data banks and neurometrics: Basic concepts, methods and results of norm construction. In Handbook of Electroencephalography and Clinical Neurophysiology, Vol. I; Gevins, A.S., Remond, A., Eds.; Elsevier: Amsterdam, The Netherlands, 1987; pp. 449–495.

[29] John ER: The neurophysics of consciousness. Brain Res. Brain Res. Rev. 2002, 39, 1–28.

[30] Szava S, Valdes P, Biscay R, Galan L, Bosch J, Clark I, Jimenez JC: High resolution quantitative EEG analysis. Brain Topogr. 1994, 6, 211–219.

[31] Hughes JR, John ER: Conventional and quantitative electroencephalography in psychiatry. J. Neuropsychiatry Clin. Neurosci. 1999, 11, 190–208.

[32] Kondacs A, Szabo M: Long-term intra-individual variability of the background EEG in normal. Clin. Neurophysiol. 1999, 110, 1708–1716.

[33] Kropotov JD. Quantitative EEG, event-related potentials and neurotherapy. Academic Press, 2009, Amsterdam.

[34] Pastor J, Vega-Zelaya L, Martin Abad E: Specific EEG encephalopathic pattern in SARS-CoV-2 patients. J. Clin. Med, 2020, 9,1545; doi:103390/ jcm9051545.

[35] Van Drongelen, W. Signal Processing for Neuroscientists; Elsevier: Amsterdam, The Netherlands, 2007.

[36] Sutter R, Kaplan PW: Clinical and electroencephalographic correlates of acute encephalopathy. J Clin Neurophysiol. 2013 Oct;30(5):443–53. doi:10.1097/WNP.0b013e3182a73bc2.

[37] Lorente L, Martín MM, González-Rivero AF, Argueso M, Ramos L, Solé-Violán J, Cáceres JJ, Jiménez A, Borreguero-León JM: Serum levels of caspase-cleaved cytokeratin-18 in patients with severe traumatic brain injury are associated with mortality: a pilot study. PLoS One 2015, 10, e0121739

[38] Lorente L, Martín MM, González-Rivero AF, Pérez-Cejas A, Argueso M, Ramos L, Solé-Violán J, Cáceres JJ, Jiménez A, García-Marín V: High Serum Caspase-Cleaved Cytokeratin-18 Levels and Mortality of Traumatic Brain Injury Patients. Brain Sci. 2019 Oct 10;9(10). pii: E269. doi: 10.3390/brainsci9100269

[39] Pandian JD, Cascino GD, So EL, Manno E, Fulgham JR: Digital video-electroencephalographic monitoring in the neurological-neurosurgical intensive care unit: clinical features and outcome. Arch Neurol 2004;61:1090–1094.

[40] Friedman D, Claassen J, Hirsch LJ: Continuous electroencephalogram monitoring in the intensive care unit. Anesth Analg 2009; 109: 506–523.

[41] Sutter R, Kaplan PW, Rüegg S: Independent external validation of the status epilepticus severity score. Crit

Care Med. 2013 Dec;41(12): e475–9. doi: 10.1097/CCM.0b013e31829eca06

[42] Alvarez V, Rossetti AO: Clinical Use of EEG in the ICU: Technical Setting. J Clin Neurophysiol. 2015 Dec;32(6): 481–5. doi: 10.1097/WNP.0000000000000194.

[43] Akman CI, Micic V, Thompson A, Riviello, JJ Jr: Seizure detection using digital trend analysis: factors affecting utility. Epilepsy Res 2011;93: 66–72.

[44] Williamson CA, Wahlster S, Shafi MM, Westover MB: Sensitivity of compressed spectral arrays for detecting seizures in acutely ill adults. Neurocrit Care 2014;20:32–39.

[45] Fisher R, van Emde Boas W, Blume W, Elger C, Genton P, Lee P, Engel J: "Epileptic seizures and epilepsy: definitions proposed by the International League Against Epilepsy (ILAE) and the International Bureau for Epilepsy (IBE)". Epilepsia. 2005; 46 (4): 470–2. doi:10.1111/j.0013-9580.2005.66104.x.

[46] Sun T, Guan J: Novel coronavirus and central nervous system. Eur J Neurol 2020 Mar 26, doi: 10.1111/ene.14227, PMID: 32216009.

[47] Wu Y, Xu X, Chen Z, Duan J, Hashimoto K, Yang L, Liu C, Yang C: Nervous system involvement after infection with COVID-19 and other coronaviruses. Brain Behav Immun 2020 Mar 30, doi: 10.1016/j.bbi.2020.03.031, PMID: 32240762.

[48] Moriguchi T, Harii N, Goto J, Harada D, Sugawara H, Takamino J, Ueno M, Sakata H, Kondo K, Myose N, Nakao A, Takeda M, Haro H, Inoue O, Suzuki-Inoue K, Kubokawa K, Ogihara S, Sasaki T, Kinouchi H, Kojin H, Ito M, Onishi H, Shimizu T, Sasaki Y, Enomoto N, Ishihara H, Furuya S, Yamamoto T, Shimada S: A first case of meningitis/encephalitis associated with SARS-Coronavirus-2. Int J Infect Dis. 2020 May;94:55–58. doi: 10.1016/j.ijid.2020.03.062. Epub 2020 Apr 3. PMID: 32251791; PMCID: PMC7195378.

[49] Mehta P, McAuley DF, Brown M, Sánchez E, Tattersall RS, Manson JJ: COVID-19: consider cytokine storm syndromes and immunosuppression. Lancet 2020, 395, 1033–1034, doi: 10.1016/S0140-6736(20)30628-0.

[50] Vaira LA, Salzano G, Deiana G, De Riu G: Anosmia and ageusia: common findings in COVID-19 patients. Laryngoscope 2020 Apr 1, doi: 10.1002/lary.28692, PMID: 32237238.

[51] Filatov A, Sharma P, Hindl F, Espinosa PS: Neurological complications of coronavirus (COVID- 19): encephalopathy. Cureus 2020, 12, e7352, doi:10.7759/cureus.7352.

[52] Poyiadji N, Shahin G, Noujaim D, Stone M, Patel S, Griffith B: COVID-19-associated Acute Hemorrhagic Necrotizing Encephalopathy: CT and MRI Features. Radiology 2020 Mar 31, doi: 10.1148/radiol.2020201187, PMID: 32228363.

[53] Chen, C.; Zhang, X.R.; Ju, Z.Y.; He, W.F. Advances in the research of cytokine storm mechanism induced by Corona Virus Disease 2019 and the corresponding immunotherapies. Zhonghua Shao Shang Za Zhi 2020 Mar 1, doi: 10.3760/cma.j.cn501120-20200224-00088, PMID: 32114747.

[54] Li Y, Fu, L, Gonzales DM, Lavi E: Coronavirus neurovirulence correlates with the ability of the virus to induce proinflammatory cytokine signals from astrocytes and microglia. J Virol 2004, 78, 3398–3406, doi: 10.1128/jvi.78.7.3398-3406.2004, PMID: 15016862.

[55] Kaplan PW, Rossetti AO: EEG patterns and imaging correlations in

encephalopathy: encephalopathy part II. J Clin Neurophysiol 2011, 28, 233–251, doi: 10.1097/WNP.0b013e31821c33a0, PMID: 21633250.

[56] Petrescu A-M, Taussig D, Bouilleret V. Electroencephalogram (EEG) in COVID-19: A systematic retrospective study. Neurophysiol Clin. 2020;50(3):155–165. doi:10.1016/j.neucli.2020.06.001

[57] Pilato MS, Urban A, Alkawadri R, Barot NV, Castellano JF, Rajasekaran V, Bagić AI, Fong-Isariyawongse JS: EEG Findings in Coronavirus Disease. J Clin Neurophysiol. 2020 Jul 1. doi: 10.1097/WNP.0000000000000752. Epub ahead of print. PMID: 32639251.

[58] Scullen T, Keen J, Mathkour M, Dumont AS, Kahn L: Coronavirus 2019 (COVID-19)-Associated Encephalopathies and Cerebrovascular Disease: The New Orleans Experience. World Neurosurg. Published online May 28, 2020. doi:10.1016/j.wneu.2020.05.192

[59] Lindau M, Jelic V, Johansson SE, Andersen C, Wahlund LO, Almkvist O: Quantitative EEG abnormalities and cognitive dysfunctions in frontotemporal dementia and Alzheimer's disease. Dement Geriatr Cogn Disord. 2003;15(2):106–14.

[60] Prichep LS: Quantitative EEG and electromagnetic brain imaging in aging and in the evolution of dementia. Ann N Y Acad Sci. 2007 Feb;1097:156–67

[61] Mizuno T, Takahashi T, Cho RY, Kikuchi M, Murata T, Takahashi K, Wada Y: Assessment of EEG dynamical complexity in Alzheimer's disease using multiscale entropy. Clin Neurophysiol. 2010 Sep;121(9):1438–1446

[62] Abásolo D, Hornero R, Gómez C, García M, López M: Analysis of EEG background activity in Alzheimer's disease patients with Lempel-Ziv complexity and central tendency measure. Med Eng Phys. 2006 May;28 (4):315–22.

[63] Abásolo D, Escudero J, Hornero R, Gómez C, Espino P: Approximate entropy and auto mutual information analysis of the electroencephalogram in Alzheimer's disease patients. Med Biol Eng Comput. 2008 Oct;46(10):1019–28.

[64] van der Zande JJ, Gouw AA, van Steenoven I, Scheltens P, Stam CJ, Lemstra AW: EEG Characteristics of Dementia With Lewy Bodies, Alzheimer's Disease and Mixed Pathology. Front Aging Neurosci. 2018 Jul 3;10:190. doi: 10.3389/fnagi.2018.00190. eCollection 2018.

[65] Szelenberger W, Skalski M, Franaszczuk P, Mitraszewski P, Blinowska K: Pharmaco-EEG analysis by means of FFT and FAD. Acta Physiol Pol. 1989 Jul-Aug;40(4):423–30.

[66] Akin M: Comparison of wavelet transform and FFT methods in the analysis of EEG signals. J Med Syst. 2002 Jun;26(3):241–7. doi: 10.1023/a:1015075101937.

[67] Höller Y, Trinka E, Kalss G, Schiepek G, Michaelis R. Correlation of EEG spectra, connectivity, and information theoretical biomarkers with psychological states in the epilepsy monitoring unit - A pilot study: Epilepsy Behav. 2019 Oct;99:106485. doi: 10.1016/j.yebeh.2019.106485. Epub 2019 Sep 4.

[68] Gast H, Schindler K, Rummel C, Herrmann US, Roth C, Hess CW, Mathis J: EEG correlation and power during maintenance of wakefulness test after sleep-deprivation. Clin Neurophysiol. 2011 Oct;122(10):2025–31. doi: 10.1016/j.clinph.2011.03.003. Epub 2011

[69] Kairui Guo, Candra H, Hairong Yu, Huiqi Li, Nguyen HT, Su SW: EEG-

based emotion classification using innovative features and combined SVM and HMM classifier. Annu Int Conf IEEE Eng Med Biol Soc. 2017 Jul;2017: 489–492. doi: 10.1109/EMBC.2017. 8036868.

[70] Sanz-García A, Pérez-Romero M, Pastor J, Sola RG, Vega-Zelaya L, Monasterio F, Torrecilla C, Vega G, Pulido P, Ortega G: Potential EEG biomarkers of sedation doses in intensive care patients unveiled by using a machine learning approach. Journal of Neural Engineering, 2019; 16: 026031. https://doi.org/10.1088/1741-2552/ ab039f.

[71] Lenartowicz A, Loo SK: Use of EEG to diagnose ADHD. Curr Psychiatry Rep. 2014 Nov;16(11):498. doi: 10.1007/ s11920-014-0498-0.

[72] Jiang MJ, Zhang HJ, Li WR, Wu WQ, Huang YM, Xu DM, Qi YY, Qin KY, Zhang L, Zhang JL: Analysis of EEG Lemple-Ziv complexity and correlative aspects before and after treatment of anti-syphilis therapy for neurosyphilis. Neurol Res. 2019 Mar;41 (3):199–203. doi: 10.1080/ 01616412.2018.1520438.

[73] Sanz-García A, Pérez-Romero M, Pastor J, Sola RG, Vega-Zelaya L, Monasterio F, Torrecilla C, Vega G, Pulido P, Ortega GJ: Identifying causal relationships between EEG activity and intracranial pressure changes in neurocritical care patients. J Neural Eng. 2018 Dec;15(6):066029. doi: 10.1088/ 1741-2552/aadeea. Epub 2018 Sep 5. PMID: 30181428.

[74] Sanz-García A, Pérez-Romero M, Pastor J, Sola RG, Vega-Zelaya L, Monasterio F, Torrecilla C, Vega G, Pulido P, Ortega G. Potential EEG biomarkers of sedation doses in intensive care patients unveiled by using a machine learning approach. Journal of Neural Engineering, 2019; 16: 026031. https://doi.org/10.1088/1741-2552/ab039f.

Chapter 7

Periodic EEG Patterns in the Intensive Care Unit (ICU): Definition, Recognition and Clinical Significance

Boulenouar Mesraoua, Musab Abdalhalim Ali, Rola Hosni Mohamed Hashem Khodair, Yazan Nofal, Dirk Theophiel O. Deleu, Hassan Jasim Al Hail, Osama Robin Elalamy, Gayane Melikyan, Covanis Athanasios and Ali Akbar Asadi-Pooya

Abstract

Periodic electroencephalographic (EEG) patterns are frequently recorded during ICU EEG monitoring in patients with altered mental status; these EEG features represent electrical discharges, ictal in appearance, occuring at regular intervals. They are known as lateralized periodic discharges (LPDs), bilateral independent periodic discharges (BIPDS), generalized periodic discharges (GPDs), continuous 2/s GPDs with triphasic morphology or triphasic waves (TWs) and Stimulus Induced Evolving Lateralized Rhytmic delta activity or Si-Evolving LRDA (previously SIRPIDS); other periodic, rhythmic patterns are Occasional frontally predominant brief 2/s GRDA (FIRDA previously), Lateralized rhythmic delta activity (LRDA) and Brief potentially ictal rhythmic discharges or B (I)RDs. The role of most (not all) of these EEG patterns is controversial; there is no consensus on which patterns are associated with ongoing seizure injury, which patterns need to be treated, and how aggressively they should be treated. Many authors consider these patterns as an unstable state on an ictal-interictal EEG continuum; the aim of the present chapter is to gain knowledge of these EEG features, show their association with known neurologic pathologies/syndromes and finally how to manage them.

Keywords: ICU, periodic EEG patterns, altered mental status, treatment

1. Introduction

Periodic electroencephalographic (EEG) patterns are frequently recorded during ICU EEG monitoring in patients with Altered Mental Status (AMS) [1]; these controversial EEG patterns consist of discharges usually epileptiform in appearance, which occur at regular intervals, in critically ill patients. They are commonly classified as periodic lateralized discharges (PLDs), bilateral independent PLDs or BIPLDs, generalized periodic discharges (GPDs) and triphasic waves. Stimulus-induced rhythmic, periodic or ictal discharges (SIRPIDs) are peculiar EEG patterns, which may be present as periodic discharges. Other periodic, rhythmic patterns are Occasional frontally predominant brief 2/s GRDA (FIRDA previously), Lateralized rhythmic delta activity (LRDA) and Brief potentially ictal rhythmic discharges or B (I)RDs.

There is still no consensus on which specific EEG features are associated with ongoing neuronal injury, which ones should be treated and how aggressively they should be treated [2]. In critically ill patients, a high index of suspicion of ongoing status epilepticus (SE), particularly non-convulsive epileptic status (NCSE) should alert the intensive care team and the neurologist, in the presence of these EEG periodic discharges, especially in patients with impaired consciousness, prompting the use of antiseizure medications [3, 4]. In addition, the occurrence of such EEG features may be in favor of cerebral impairment, acute or subacute [4, 5].

The aim of this study is to make a review of these periodic EEG features, emphasizing the importance of their recognition and clinical significance. Their clinical significance is uncertain, it is related to a variety of etiologies, and many authors suggest that these patterns are unequivocally epileptogenic in some cases. Their recognition and classification are important to establish a correlation between clinical, neurological, neuroimaging data with the EEG results.

2. Historical note

There is still a scientific debate regarding these EEG features. Until recently, there was no uniformly accepted nomenclature for those frequently encountered ICU EEG abnormalities such as periodic epileptiform discharges, fluctuating rhythmic patterns and combinations.

Based on these questions, the American Clinical Neurophysiology Society (ACNS) recently proposed new terminology for these controversial EEG patterns [6, 7] (**Table 1**); terms such as "triphasic waves," which implies for many clinicians a metabolic encephalopathy was eliminated; in addition, the use of "ictal," "interictal"

New Terms for Older Terms	
OLD Term	**NEW Term**
Triphasic waves, most of record	**= continuous 2/s GPDs (with triphasic morphology)**
PLEDs	**= LPDs**
BIPLEDs	**= BIPDs**
GPEDs/PEDs	**= GPDs**
FIRDA	**= Occasional frontally predominant brief 2/s GRDA (if 1–10% of record)**
PLEDS+	**=LPDs+**
SIRPIDs* w/ focal evolving RDA	**=SI-Evolving LRDA**
Lateralized seizure, delta frequency	**= Evolving LRDA**
Semirhythmic delta	**= Quasi-RDA**
SIRPIDs: stimulus-induced rhythmic, periodic or ictal discharges	

PLEDS: Periodic Lateralized Epileptiform Discharges; LPDs: Lateralized Periodic Discharges; BIPLEDs: Bilateral Independant Periodic Lateralized Epileptiform Discharges; BIPDs: Bilateral Independant Periodic Discharges; GPEDs/PEDs: Generalized Periodic Epileptiform Discharges/Periodic Epileptiform Discharges; FIRDA: Frontal Intermitent Rhytmic Delta Activity; LRDA: Lateralized rhythmic Delta Activity; SIRPIDS: Stimulus-induced rhythmic, periodic or ictal discharges

Table 1.
ICU periodic EEG patterns: Old vs. new terminology [6].

and "epileptiform" for the equivocal patterns was ovoided; these EEG features are now termed Periodic Lateralized Discharges (PLDs), (formerly Periodic lateralized Epileptiform Discharges or PLEDs), Bilateral independent Periodic Discharges or BIPLDs, (formerly Bilateral independent Periodic Lateralized Epileptiform Discharges or BIPLEDs), Generalized Periodic Discharges (GPDs), formerly Generalized Periodic Epileptiform Discharges (GPEDs) and continuous GPD 2 / s with triphasic morphology) (formerly Triphasic Waves or TWs). SI-Evolving LRDA, Stimulus-induced Lateralized Rhythmic Delta Activity (formerly stimulus-induced rhythmic, periodic or ictal discharges or SIRPIDs) are also peculiar, recently discovered EEG patterns, and may manifest as periodic discharges [8]. The other periodic and rhythmic patterns are the occasional brief 2 / s GRDA predominantly frontal (formerly Frontal Intermitent Rhythmic Delta Activity or FIRDA), Evolving Lateralized Rhythmic Delta Activity or Evolving LRDA, (formerly Lateralized Rhythmic Delta Activity (LRDA) or Lateralized seizure delta activity, and Short ictal potential rhythmic discharges or B(I)RDs, will also be briefly discussed in this chapter.

3. Lateralized periodic discharges (LPDs)

3.1 EEG characteristics

LPDs are stereotyped, repetitive EEG discharges and recurr periodically at regular intervals at 0.5 to 3 Hz; they are broadly lateralized over one hemisphere, particularly over the parasagittal and temporal areas; LPDs are usually epileptiform in appearance; they appear like sharp waves/sharp waves complexes ranging from 50 to 300 uV in amplitude or as blunt delta waves that recur in stereotyped periodic fashion (**Figure 1**). They are maximal in any focal brain lesion, sometimes asymmetrical (**Figure 2**). They are also associated with additional EEG evidence of ipsilateral cerebral dysfunction such as focal slowing, loss of posterior dominant rhythm.

3.2 Frequency

LPDs frequency on continuous EEG Monitoring (cEEG) varies from 6.2% to 8.6% [9]. In the intensive care unit, LPDs are found in 47% of patients [9]. LPDs are most commonly seen in patients with focal neurological deficit and are associated with varying degree of altered consciousness.

3.3 Etiology

LPDs are frequently associated with acute, structural brain lesion; very often with Ischemic stroke, viral encephalitis, including autoimmine encephalitis [10], tumors, intracerebral hemorrage (ICH) [11], Anoxic encephalopathy, Creutzfeld-jacob disease CJD [12], Subarachnoid Hemorrage (SAH) [13], Multiple Sclerosis (MS), Posterior Reversible Encephalopathy Syndrome (PRES) (**Figure 3**); they have been also reported in Migraine headache, Mitochondrial Encephalopathy with lactic acidosis and stroke like episodes, (MELAS).

LPDs are most commonly associated with cortical gray matter or subcortical gray and white matter lesion [14]; however no structural abnormality is found on neuroimaging in 25–33% of patients with LPDs.

3.4 Significance

Are LPDs a transient EEG phenomenon following acute neurologic insult resolving usually within days to weeks? or a chronic phenomenon associated with

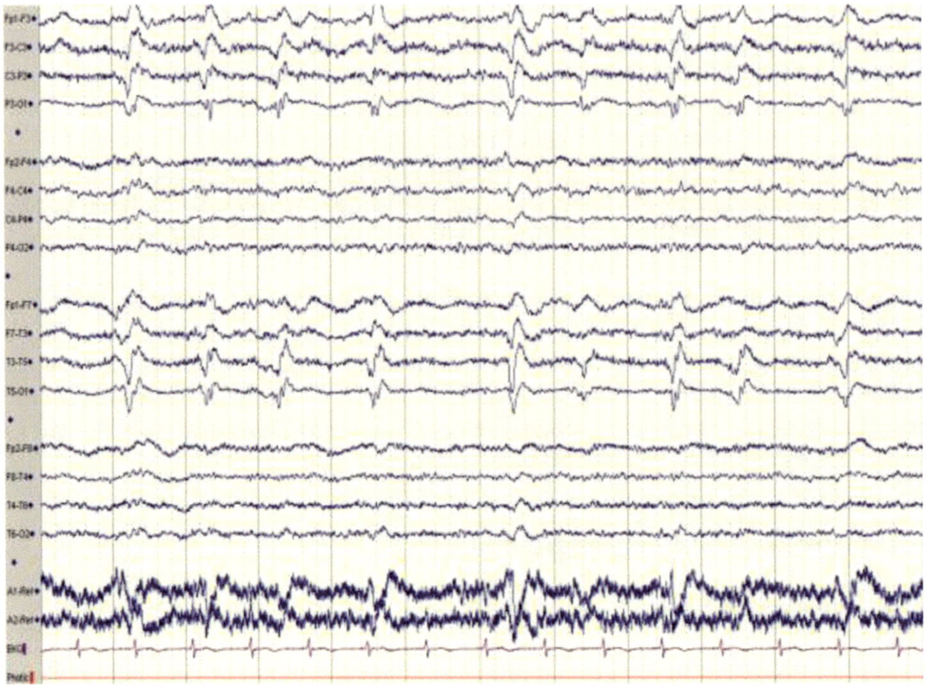

Figure 1.
LPDs in a 70ys old male patient with HTN. Brain MRI compatible with PRES syndrome. Quasi-periodic LPDs lateralized over the left hemisphere. PRES: Posterior Reversible Encephalopathy Syndrome.

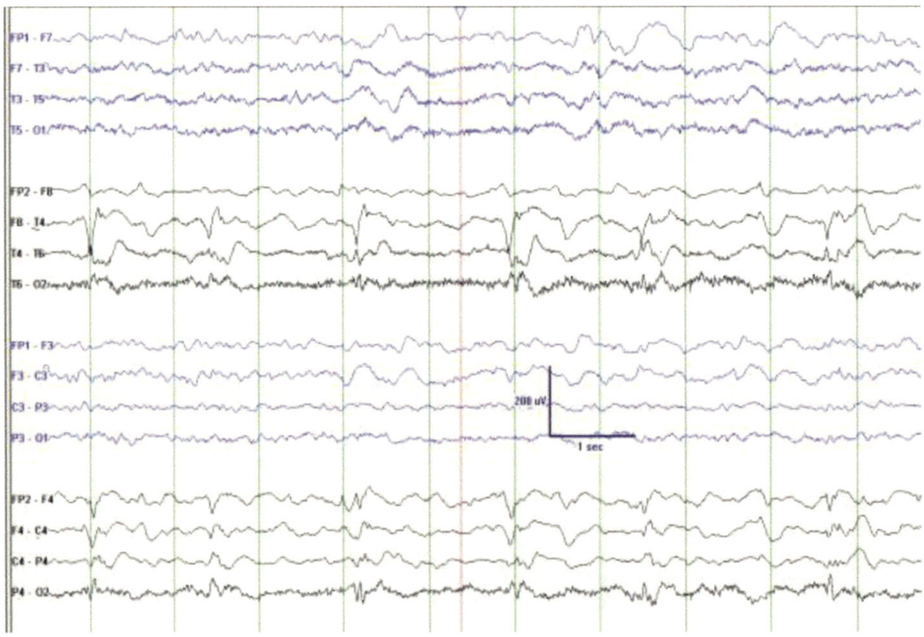

Figure 2.
Unilateral, LPDs in a 61 ys old patient with right temporal hemorrhage.

epilepsy? Studies have reported that LPDs are found in 5–30% of patients with history of epilepsy [15], in 26% of patients with remote brain injury and epilepsy [16] and also in patients with structural brain lesions and symptomatic epilepsy [17];

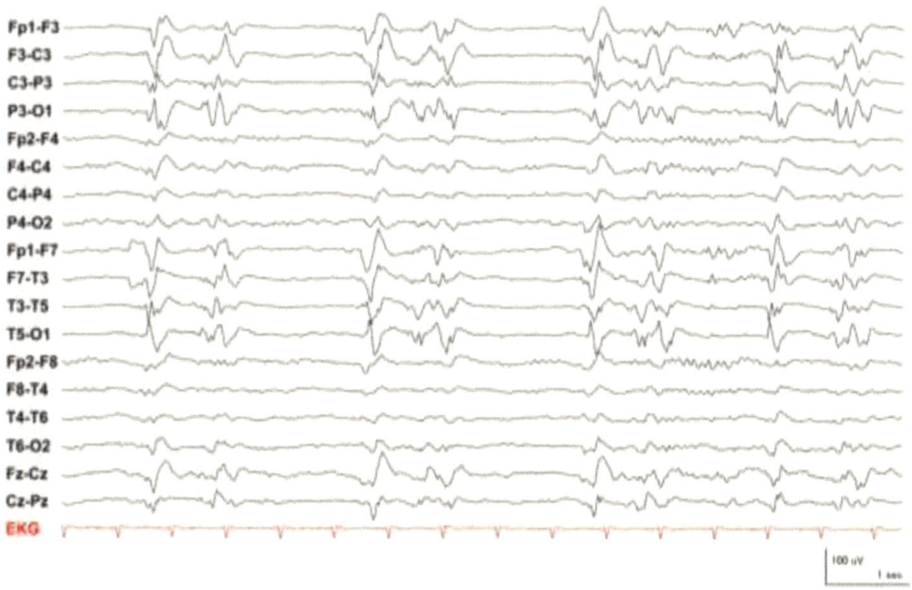

Figure 3.
"LPDs Plus" with complex morphology and prolonged after discharges; "Ictal appearing" LPDs in a 50 ys old patient with left sided stroke and right sided clonic seizures.

some authors consider LPDs as an unstable, potentially epileptogenic state, a pattern on the "ictal-interictal continuum [18].

3.5 LPDs and seizures: LPDs ictal pattern??

The presence of LPDs in patients with altered mental status (AMS) is associated with increased risk of seizures. Clinical seizures are indeed very frequent in patients with LPDs; focal motor seizures are the most common (**Figure 3**) [19]; such seizures can occur prior or at the same time as LPDs [20]; the risk of developing subsequent seizures following LPDs is 10–56% [21]; LPDs may represent an ictal pattern when associated with clinical correlate such as focal clonic seizures (Stroke) and Epilepsia Partialis continua (EPC); LPDs may be ictal when associated with subtle clinical manifestations such as eye deviation, aphasia, hemianopsia in patients with AMS; in this setting both LPDs and clinical symptoms improve with antiseizure drugs (ASD); Claassen et al. have reported that LPDs are highly associated with Non Convulsive Seizures (NCSs), as high as 40% [22]. The frequency of LPDs is correlated with seizure risk [23]: LPDs of less than 1 Hz: 40% risk of seizures, LPDs of 2 Hz or greater: 66% risk of seizures. "Lateralized Periodic Discharges Plus ("LPDs plus") are LPDs with a Complex morphology, a prolonged after discharges and an "Ictal appearing" (**Figures 3–5**); in addition, intervening fast activities (LPD + F) (**Figures 3** and **5**), superimposed rhythmic activity (LPD + R) or both (LPD + FR) can complicate this picture. "LPDs plus" have a rapid repetitive rate (>2HZ) and are highly associated with clinical seizures [24].

3.6 Management of patients with LPDs

There are no clear data regarding the management of the LPDs in patients with AMS. However neuroimaging should be performed in all patients with LPDs; metabolic/reversible conditions should be treated; as mentioned above,

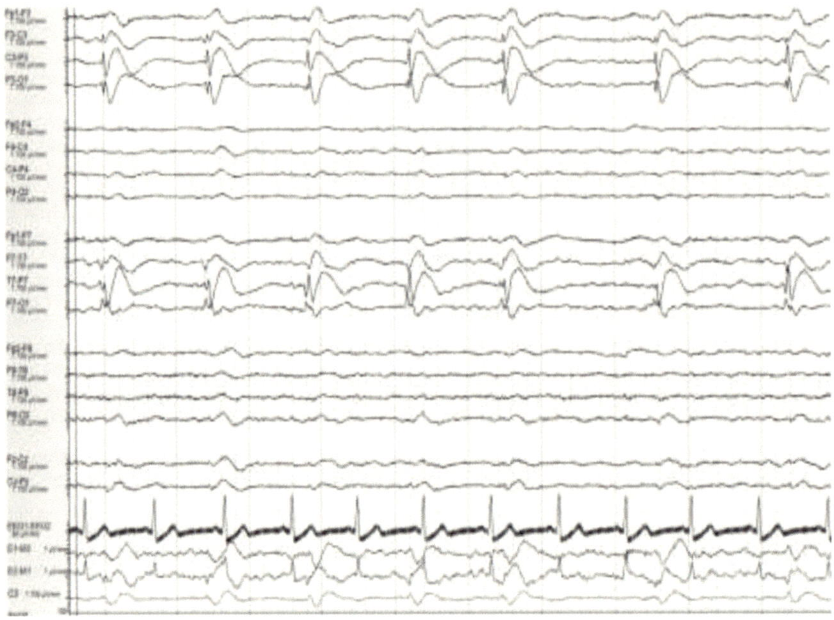

Figure 4.
LPDs at a frequency of 0.5 to 1/s with a spiky appearence running at nearly regular interval.

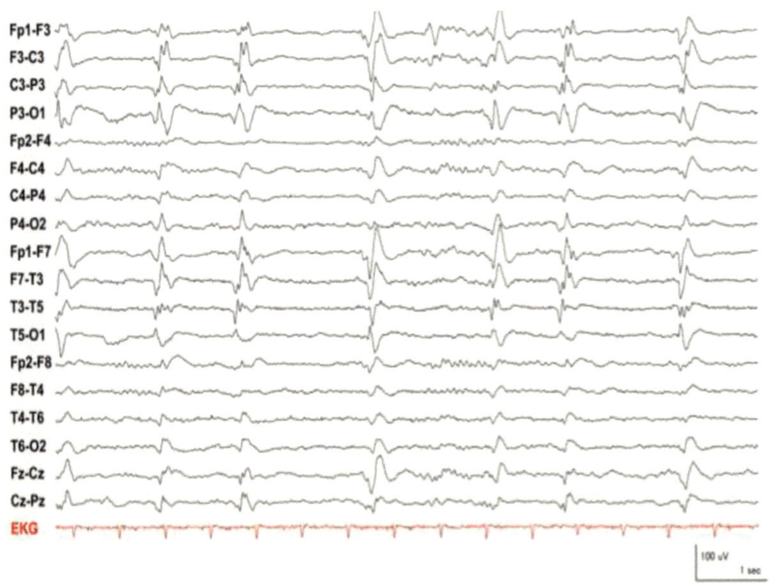

Figure 5.
64 year old left MCA infarction with jerky movements of the right upper limb. MCA: Middle Cerebral Artery.

prolonged EEG monitoring (>24 h) is recommended in the presence of LPDs because of their association with seizures, particularly NCSs and NCSE. When LPDs are found in a confused patient, a benzodiazepine trial, such as lorazepam IV should be considered and the patient monitored (**Figures 6** and 7). The clinical significance and management of LPDs in comatose patients is controversial [25] and there no available data regarding the continuation of ASDs after hospitalization.

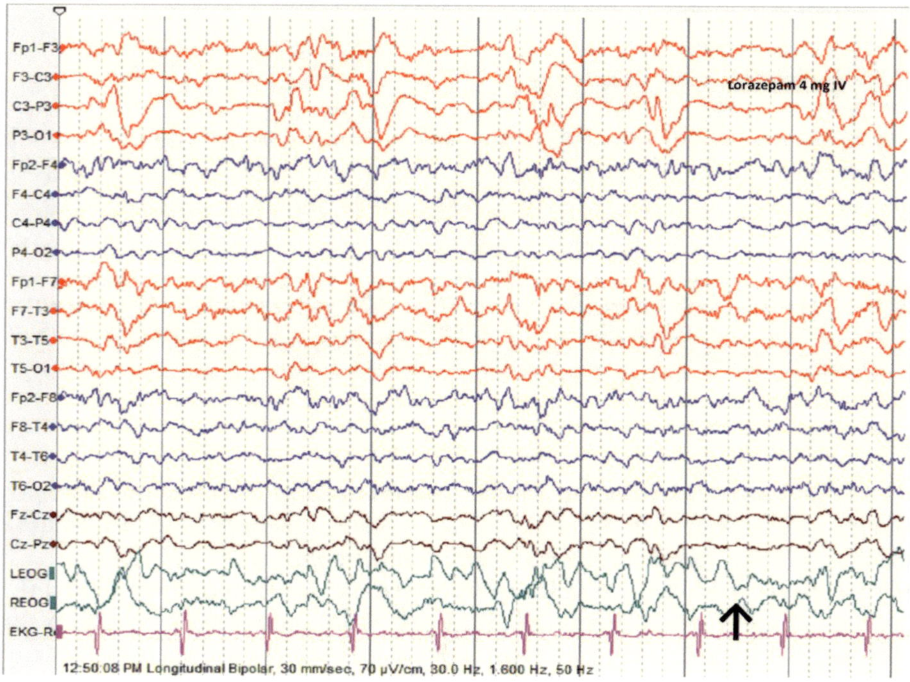

(a)

Figure 6.
LPDs in a 58 ys old male admitted to ICU with AMS, confused; no abnormal movements; given lorazepam 4 mg IV. AMS: Altered mental status; ICU: Intensive care unit.

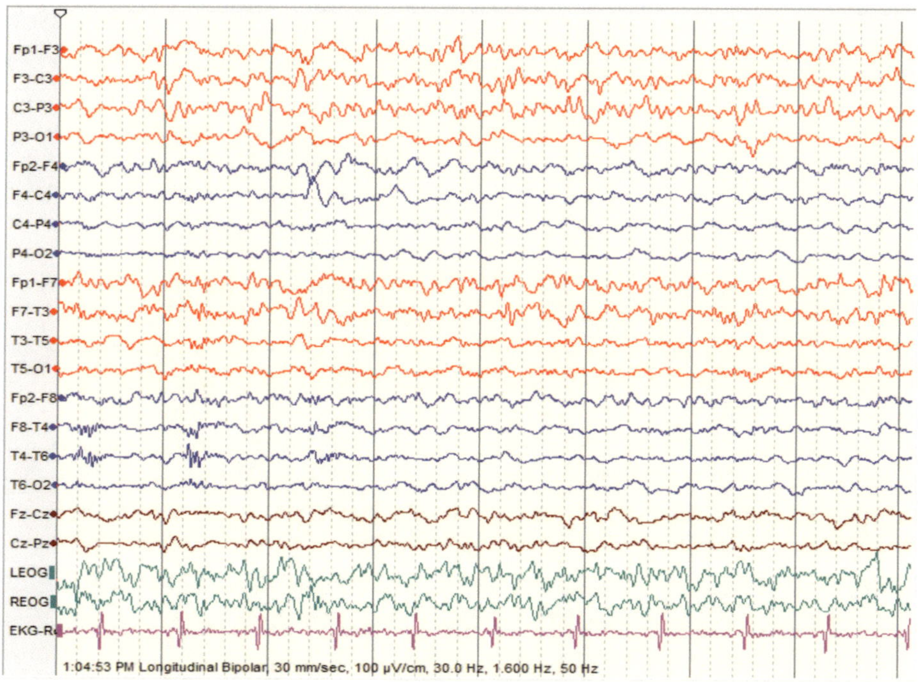

(b)

Figure 7.
*Same patient in **Figure 6,** 14mn following lorazepam IV. Note the dramatic EEG improvement; the patient also showed an improvement in the level of consciousness.*

4. Generalized periodic discharges (GPDs)

4.1 EEG characteristics, frequency

GPDs are symmetric, bilateral, repetitive discharges with regular morphology and a frontal or occipital predominance; they show the same morphology and in general the same interdischarge interval (quasi-periodic in more than 50%); slowing of the delta or theta range, frequently a background suppression of cortical activity accompagny GPDs (**Figure 8**). Their occurrence with LPDs and BIPDs in the same patient is not rare [26]. GPDs are frequently encountered in comatose patients (>55%) and very early during cEEG Monitoring and are observed in 5–10% during cEEG [26].

4.2 Etiology

GDPs have been described in hypoxic anoxic encephalopathy, acute ischemic stroke, herpes encephalitis, Sepsis, Systemic lupus Erythematous (SLE), traumatic brain injury (TBI), hepatic encephalopathy, Hashimoto's encephalitis, hypo/ hypernatremia, uremia, renal failure, hypoglycemia, hypothyroidism, Epileptic Encephalopathy, Status Epilepticus (SE), Creurzfeld Jacob disease, Steroid responsive encephalopathy Subacute, Sclerosing Panencephalitis (SSPE), Alzheimer Disease, Benzodiazepines, Barbiturates, propofol withdrawal; with baclofen, lithium, Phencycline or ketamine, Tiagabine and Cyclosporine, Cefepime, and other cephalosporines [27–31].

4.3 GPDs and seizures

GPDs are associated with seizures [32]; one study found NCSs in 27% vs. 8% of patients with/without GPDs and NCSE in 22% vs. 7% only [26]. GPDs with "plus

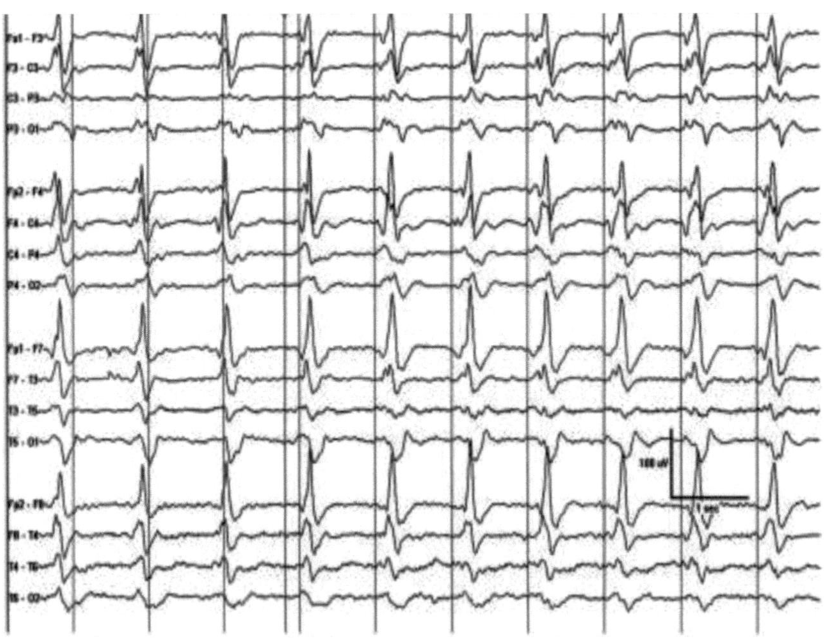

Figure 8.
GPDs. One per second GDPs with clear frontal predominance and a sharp morphology.

DOI: http://dx.doi.org/10.5772/intechopen.95503

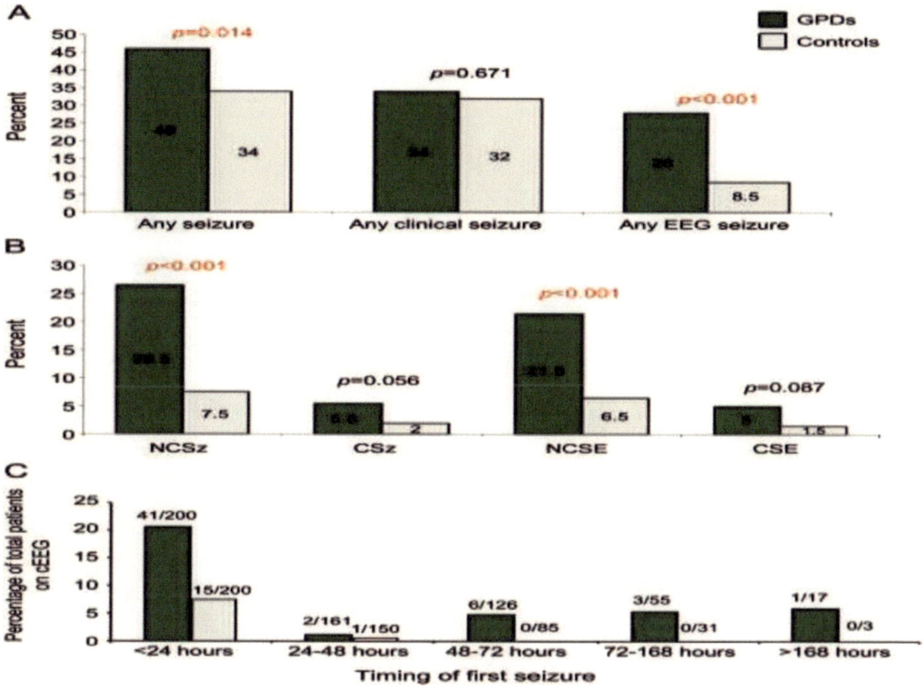

Figure 9.
Association of NCSs, NCSE with GDPs. Generalized periodic discharges and NCSs, NCSE. (A) Seizure occurrence at any time in patients with GPDs vs. controls (%) and (B) during cEEG in patients with GPDs vs. controls (%). (C) Timing of first recorded seizure in patients with GPDs vs. controls. CSE convulsive status epilepticus; CSz convulsive seizure; NCSE nonconvulsive status epilepticus; NCSz nonconvulsive seizure.

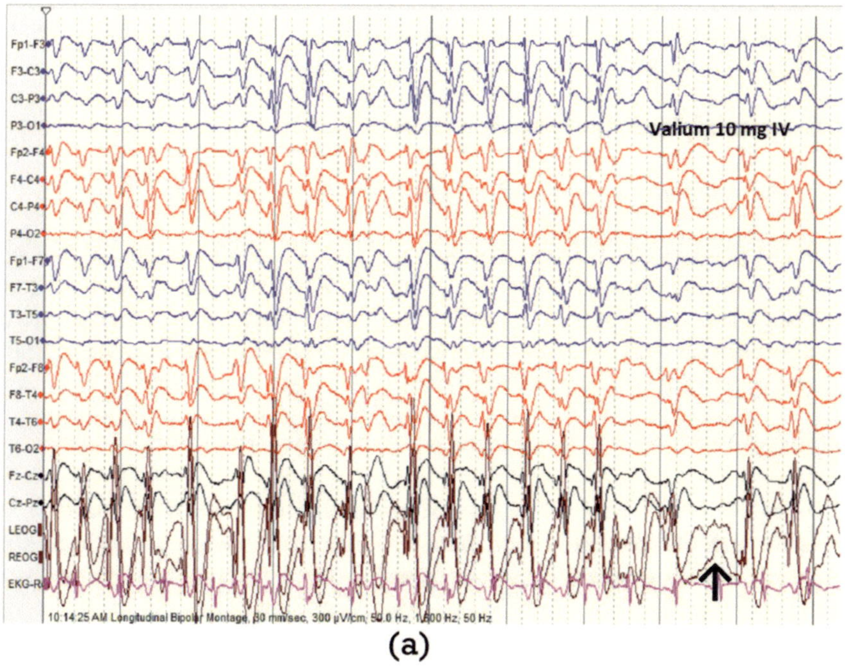

Figure 10.
GPDs with Triphasic morphology (GDPs TM) in a 59 ys old patient following cardiac arrest. GDPs show frequency change and qualify probably for "evolving GPDs". Patient given 10 mg diazepam IV.

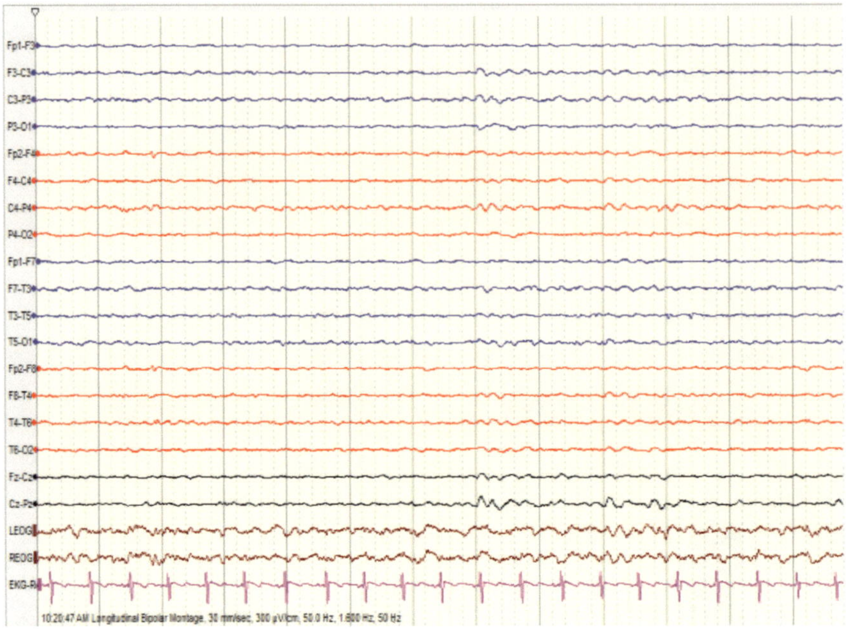

(b)

Figure 11.
*Same patient in **Figure 10** following 10 mg of diazepam. Note the EEG improvement 6 mn following diazepam IV; however, there was no clinical improvement (patient remained comatose): "Possible NCSE in coma" ("comatose NCSE"?).*

features", sharper morphology, and high frequency increase the possibility of seizures on EEG [23]. Sutter R et al. have shown that GDPs are strongly associated with NCSs, NCSE [33] (**Figure 9**). GDPs may reflect an ictal rhythm; they may manifest with a triphasic morphology [31]; in addition there is a clear GDPs response (triphasic morphology) to Benzodiazepines (BDZs) (**Figures 10** and **11**).

4.4 GPDs with triphasic morphology (GDPs TM)

4.4.1 EEG characteristics

GPDs TM (**Figure 10**) [7] are known as moderate to high voltage EEG discharges (100-300uv), with a frequency varying from 1 to 2 Hz; they are usually characterized by the presence of 3 phases: Negative–positive –negative in polarity with a predominant positive phase; often, there is a frontal predominance, and sometimes an anterior –posterior or posterior –anterior phase lag.

4.4.2 Etiology

GPDs with triphasic morphology are reported in metabolic disturbances, hypertensive encephalopathy, subcortical white-matter disease, infections, strokes, hypoglycemia hypernatremia/hyponatremia, postictal states, lithium and baclofen toxicity.

4.4.3 GPDs with triphasic morphlogy and seizures

GDPs TM association with seizures has been discussed by several authors especially concerning nonconvulsive status epilepticus (NCSE) where the improvement of both EEG and patient level of consciousness with BDZs is consistent with NCSE; Jirsh et al. define a possible NCSE in a patient, a state where BDZs abolish GPDs TM without clinical improvement [34] (**Figures 10** and **11**); patients with GDPs and TM are likely to develop seizures as those without (25% vs. 26%) [35]; furthermore anterior to posterior "phase lag" does not occur with GDPs TWs related NCSE.

4.4.4 Significance, prognosis of GDPs

In one study the mortality of patients with GPDs is estimated at 53% [36]. Predictor of poor outcome in patients with GDPs include Dementia, altered to poor mental status, focal neuroimaging abnormalities and cardiac arrest [36]. Few Patients with GPDs and anoxic brain damage regain consciousness [37].

5. Bilateral independent periodic discharges BIPDs

BIPDs occur as LPDs involving both hemispheres, in an independent and asynchronous manner (**Figure 12**); The true prevalence and incidence of BIPDs is unknown; however, they are less common than LPDs; studies mention that seizures are also less common in association with BIPDs than with LPDs, successively 29% vs. 44% [23] and 43% vs. 70% [17]. BIPDs are associated with acute brain lesions (anoxic brain damage, stroke, CNS infection, tumor); in general the clinical status and prognosis of patients showing BIPDs is worse than those with LPDs; mortality rate for BIPDs compared to LPDs is much higher 47, 8% vs. 14% respectively [38].

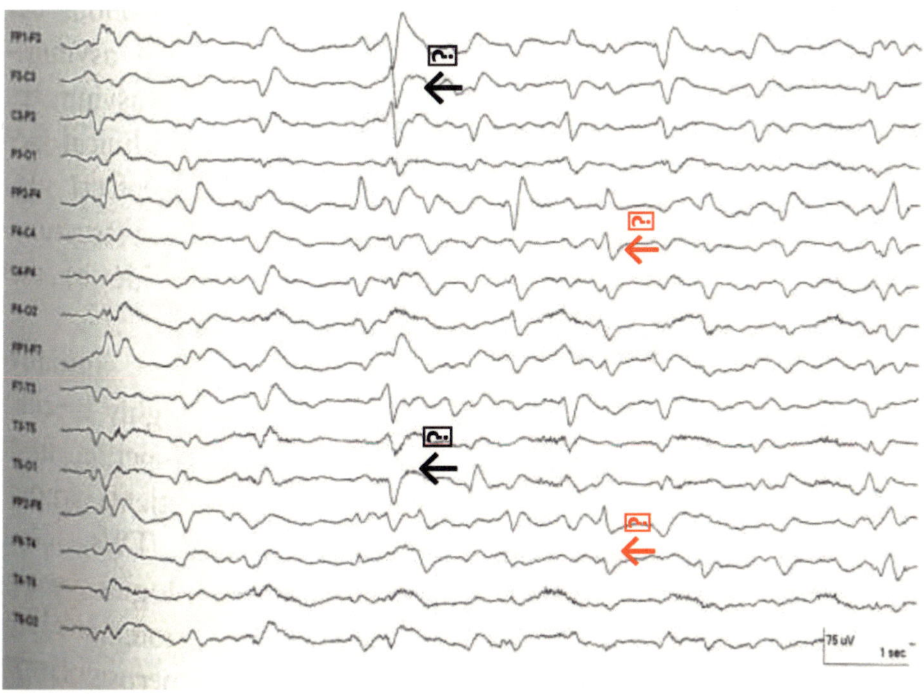

Figure 12.
Bilateral independent periodic discharges BIPDs. Black arrows show discharge arising from the left hemisphere, red arrows discharges arising from the right hemisphere.

6. Stimulus induced evolving lateralized rhytmic delta activity (Si-evolving LRDA)

These EEG features were recently described by Hirsch et al. in ICU patients [8]. They occur whenever ICU patients with AMS are stimulated using auditory, sternal rub, suctioning, or other stimulating procedures (**Figure 13**); these curious EEG patterns appear ictal, they are sometimes associated with an evolving lateralized rhythmic delta activity, stimulus induced also; Si-Evolving LRDA looks like an epileptiform activity recurring at regular intervals; there are in general purely EEG changes; however, few patients showing these features present focal motor seizures.

No data are available regarding the pathophysiology and prognostic significance of these stimuli–induced EEG features.

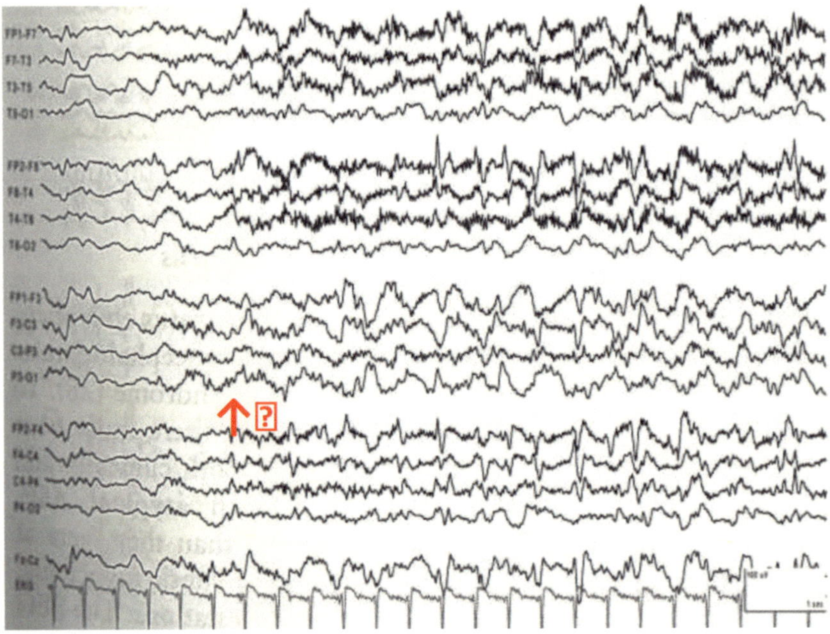

Figure 13.
Stimulus induced evolving lateralized Rhytmic delta activity (Si-evolving LRDA) in a patient with SAH. Sternal Rub (red arrow) inducing right sided periodic spikes at 2 Hz. Also note the presence of rhythmic delta activity over the left side. SAH: subarachnoid hemorrhage.

7. Occasional frontally predominant brief 2/s generalized rythmic delta activity (GRDA)

These EEG waves are bilateral, bisynchronous, symmetric activity of 4 Hz or less, intermittent, predominant anterior activities (**Figure 14**). It is the most EEG feature frequently seen in the ICU [39].

It is reported in various diseases and syndromes such as structural brain lesion, metabolic encephalopathy, epilepsy, neurodegenerative disorders.

Study has shown that this pattern is not associated with an increasing risk of seizures [23]. In general, this EEG feature represents a benign pattern and is associated with a good outcome.

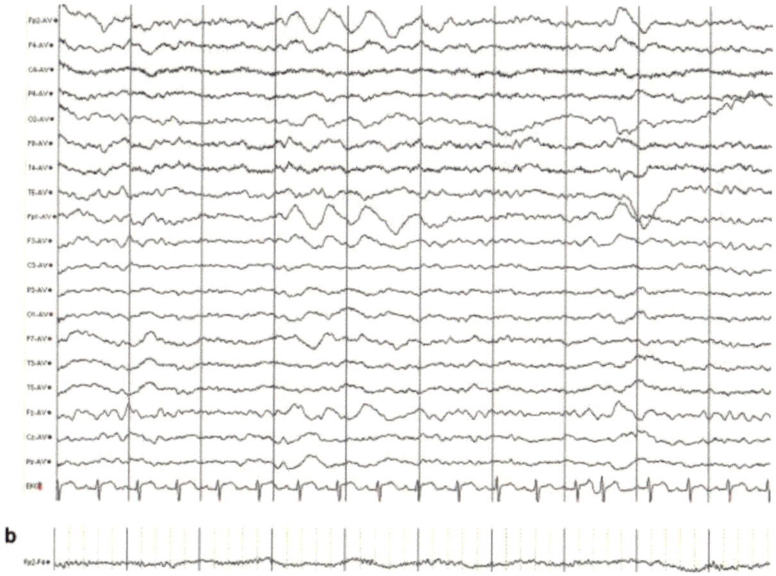

Figure 14.
Generalized Rythmic Delta activity (GRDA). Average montage. Note the brief (lasting 2 s), 2 cps, bilateral, predominantly anterior delta activity.

8. Evolving lateralized rhythmic delta activity (LRDA)

These interesting EEG features are unilateral rhythmic activity of 4 Hz or less, are found in various central nervous system lesions such as cortical and deep gray matter lesions, acute brain injury, and in chronic seizure disorder (**Figure 15**); there

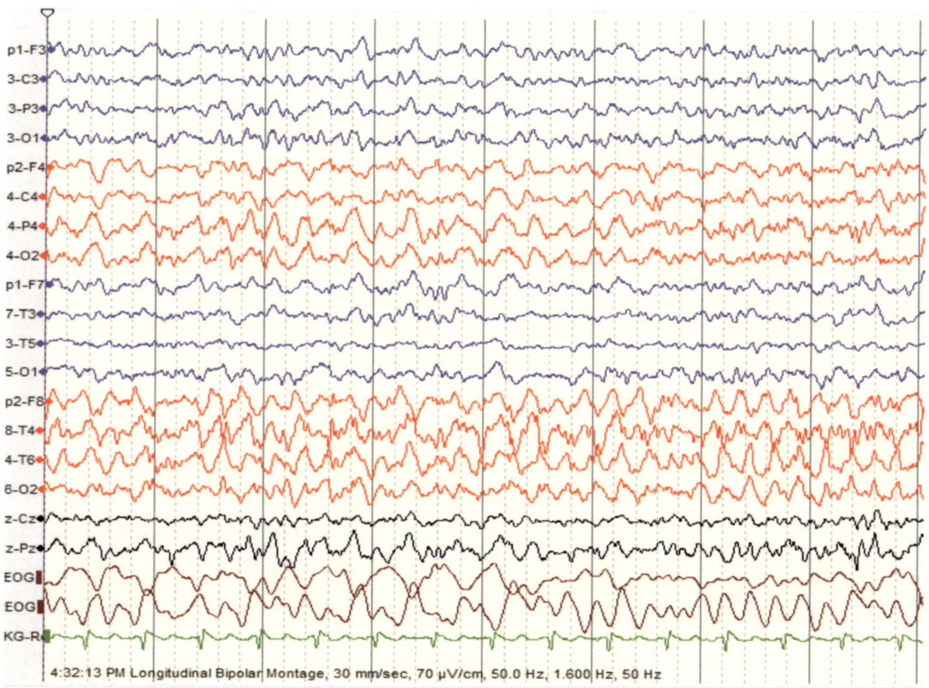

Figure 15.
LRDA in a 50 years old patient with NCSE. Note the right sided continuous lateralized focal predominantly fronto-temporal 3, 5 to 4 cps delta activity.

are also seen in acute seizures in 63% of patients [40] and are frequently associated with LPDs; LRDA carries similar implications as LPDs [40].

9. Brief potentially ictal rhythmic discharges [B(I)RDs]

B(I)RDs are very brief EEG discharges, of moderate to low amplitude, lasting between 10s and 4 s, firing at around 4 Hz or more, rhythmic focal theta activity (**Figure 16**); they are most often seen in neonates and occur in 2% in ICU population during cEEG monitoring.

Patient with B(I)RDs suffer from various neurological disorders especially chronic seizure disorder, and acute cerebral injury; a large majority of patients with B(I)RDs experience seizures.

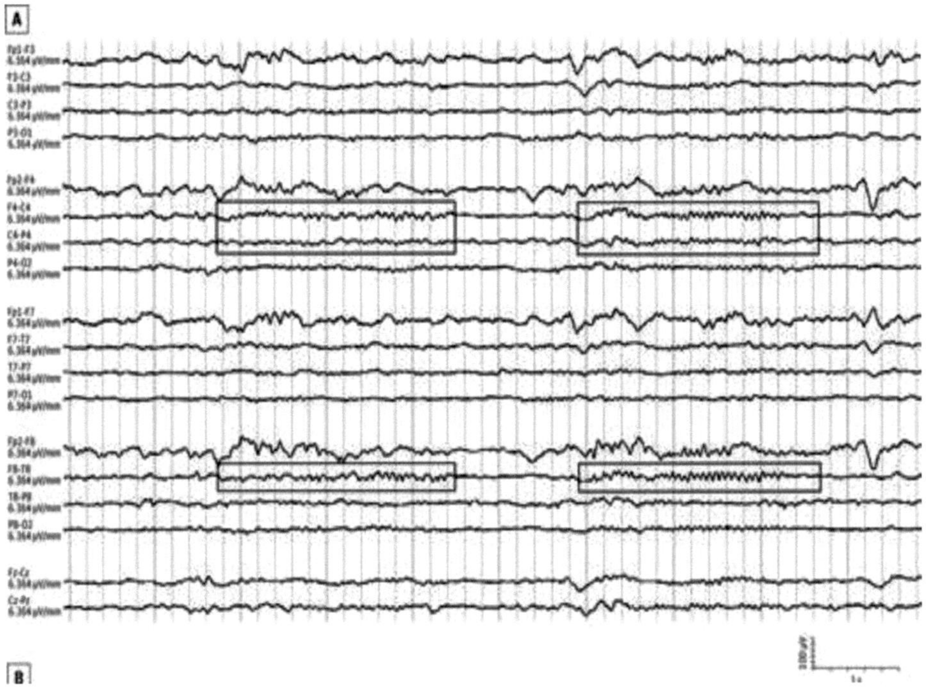

Figure 16.
B(I)RDs in a 80 ys old patient. Patient suffered from generalized tonic–clonic seizure and was unconscious. EEG shows a very brief run of sharply contoured, intermittent rhythmic theta activity over the right anterior region (boxed areas).

10. Conclusion

The electrographic patterns discussed above, such as LPDs, GPDs, LRDA, and B(I)RDs are strongly associated with seizures; this association is well established in ICU patients suffering from AMS. There are however many unanswered questions regarding these abnormal EEG features: do they reflect the severity of the brain injury? Do they actively contribute to its damage and how should we treat them are still largely unknown. Equally not established is their association with long term seizures and the risk of epilepsy.

Author details

Boulenouar Mesraoua[1,2*], Musab Abdalhalim Ali[1],
Rola Hosni Mohamed Hashem Khodair[1], Yazan Nofal[1],
Dirk Theophiel O. Deleu[1,2], Hassan Jasim Al Hail[1,2], Osama Robin Elalamy[1,2],
Gayane Melikyan[1,2], Covanis Athanasios[3] and Ali Akbar Asadi-Pooya[4,5]

1 Hamad Medical Corporation, Doha, Qatar

2 Weill Cornell Medical College, Doha, Qatar

3 Neurology and Neurophysiology Department, The Children's Hospital Agia Sophia, Athens, Greece

4 Department of Neurology, Thomas Jefferson University, Philadelphia, PA, USA

5 Department of Neurology, Shiraz University of Medical Sciences, Shiraz, Iran

*Address all correspondence to: boulenouar.mesraoua@wanadoo.fr

References

[1] Boulenouar Mesraoua, Dirk Deleu, Hassan Al Hail, Faisal Ibrahim, Gayane Melikyan, Hassan Al Hussein, Rajvir Singh, Basim Uthman, Leopold Streletz, Peter W Kaplan, Heinz Gregor Wieser. Journal of Drug Assessment 2017, https://doi.org/10.1080/21556660.2017.1396992

[2] Chong DJ, Hirsch LJ. Which EEG patterns warrant treatment in the critically ill? Reviewing the evidence for treatment of periodic epileptiform discharges and related patterns. J Clin Neurophysiol 2005;22:79-91.

[3] Hirsch LJ, Brenner RP. Periodic discharges and other controversial EEG patterns. In: Hirsch LJ, Brenner RP (Eds). Atlas of EEG in critical care. Chichester: Wiley-Blackwell, 2010: 129-160.

[4] Garzon E, Fernandes RM, Sakamoto AC. Serial EEG during human status epilepticus: evidence for PLED as an ictal pattern. Neurology 2001;57:1175-1183. [Links]

[5] Rossetti AO, Hirsch LJ, Drislane FW. Nonconvulsive seizures and nonconvulsive status epilepticus in the neuro ICU should or should not be treated aggressively: A debate.Clin Neurophysiol Pract. 2019 Aug 9;4:170-177. doi: 10.1016/j.cnp.2019.07.001. eCollection 2019.PMID: 31886441

[6] Hirsch LJ, Brenner RP, Drislane FW, et al. The ACNS subcommittee on research terminology for continuous EEG monitoring: proposed standardized terminology for rhythmic and periodic EEG patterns encountered in critically ill patients. J Clin Neurophysiol 2005;22:128-135.

[7] Hirsch LJ, LaRoche SM, Gaspard N, Gerard E, Svoronos A, Herman ST, Mani R, Arif H, Jette N, Minazad Y, Kerrigan JF, Vespa P, Hantus S, Claassen J, Young GB, So E, Kaplan PW, Nuwer MR, Fountain NB, Drislane FW. American Clinical Neurophysiology Society's Standardized Critical Care EEG Terminology: 2012 version. Journal of clinical neurophysiology: official publication of the American Electroencephalographic Society. 2013 Feb;30(1):1-27.

[8] Hirsch LJ, Claassen J, Mayer SA, Emerson RG. Stimulus-induced rhythmic, periodic, or ictal discharges (SIRPIDs): a common EEG phenomenon in the critically ill. Epilepsia 2004;45:109-123.

[9] Rai V, Jetli S, Rai N, Padma MV, Tripathi M. Continuous EEG predictors of outcome in patients with altered sensorium. Seizure. 2013 Oct;22(8):656-61. doi: 10.1016/j.seizure.2013.05.002. Epub 2013 Jun 12.PMID: 23768540

[10] Francisco Hernández-Fernández, Eva Fernández-Díaz, Jose M Pardal-Fernández, Tomás Segura, Jorge García-García Periodic lateralized epileptiform discharges as manifestation of pneumococcal meningoencephalitis.Int Arch Med. 2011; 4: 23. Published online 2011 Jun 25. doi: 10.1186/1755-7682-4-23 PMCID: PMC3141320

[11] Claassen J, Jette N, Chum F, et al. Electrographic seizures and periodic discharges after intracerebral hemorrhage. Neurology 2007; 69: 1356-1365 [PubMed] [Google Scholar]

[12] Binelli S, Agazzi P, Giaccone G, Will RG, Bugiani O, Franceschetti S, Tagliavini F. Periodic electroencephalogram complexes in a patient with variant Creutzfeldt-Jakob disease. Ann Neurol. 2006 Feb; 59(2):423-7.

[13] Claassen J, Hirsch LJ, Frontera JA, et al. Prognostic significance of continuous EEG monitoring in

patients with poor-grade subarachnoid hemorrhage. Neurocrit Care 2006; 4: 103-112 [PubMed] [Google Scholar]

[14] Lu Lin, Frank W Drislane. Lateralized Periodic Discharges: A Literature Review. J Clin Neurophysiol. 2018 May;35(3):189-198. doi: 10.1097/ WNP.0000000000000448.

[15] García-Morales I, García MT, Galán-Dávila L, Gómez-Escalonilla C, Saiz-Díaz R, Martínez-Salio A, de la Peña P, Tejerina JA. Periodic lateralized epileptiform discharges: etiology, clinical aspects, seizures, and evolution in 130 patients. J Clin Neurophysiol. 2002 Apr;19(2):172-7. doi: 10.1097/00004691-200203000-00009.PMID: 11997729

[16] Rup K Sainju, Louis N Manganas, Emily J Gilmore, Ognen A Petroff, Nishi Rampal, Lawrence J Hirsch, Nicolas Gaspard.Clinical Correlates and Prognostic Significance of Lateralized Periodic Discharges in Patients Without Acute or Progressive Brain Injury: A Case-Control Study. J Clin Neurophysiol. 2015 Dec;32(6):495-500. doi:10.1097/ WNP.000000 0000000206.

[17] Orta DS, Chiappa KH, Quiroz AZ, Costello DJ, Cole AJ Prognostic implications of periodic epileptiform discharges..Arch Neurol. 2009 Aug;66(8):985-91. doi: 10.1001/ archneurol.2009.137.PMID: 19667220

[18] Gregory Kapinos, Eugen Trinka, Peter W Kaplan. Multimodal Approach to Decision to Treat Critically Ill Patients With Periodic or Rhythmic Patterns Using an Ictal-Interictal Continuum Spectral Severity Score. J Clin Neurophysiol. 2018 Jul;35(4):314-324. doi: 10.1097/WNP.0000000000000468.

[19] Chatrian GE, Shaw CM, Leffman H. The Significance of Periodic Lateralized Epileptiform Discharges in Eeg: an Electrographic, Clinical and Pathological Study. Electroencephalogr Clin Neurophysiol. 1964 Aug;17:177-93. doi: 10.1016/0013-4694(64)90149-x. PMID: 14204927

[20] S M Snodgrass, K Tsuburaya, C Ajmone-Marsan. Clinical significance of periodic lateralized epileptiform discharges: relationship with status epilepticus. J Clin Neurophysiol. 1989 Apr;6(2):159-72. doi: 10.1097/00004691-198904000-00003.

[21] J M Walsh, R P Brenner. Periodic lateralized epileptiform discharges--long-term outcome in adults. Epilepsia. Sep-Oct 1987;28(5):533-6. doi: 10.1111/ j.1528-1157.1987.tb03684.x.

[22] Claassen J, Mayer SA, Kowalski RG, Emerson RG, Hirsch LJ. Detection of electrographic seizures with continuous EEG monitoring in critically ill patients. Neurology. 2004 May 25;62(10):1743-8. doi: 10.1212/01. wnl.0000125184.88797.62.PMID: 15159471 Review.

[23] Andres Rodriguez Ruiz, Jan Vlachy, Jong Woo Lee, Emily J Gilmore, Turgay Ayer, Hiba Arif Haider, Nicolas Gaspard, J Andrew Ehrenberg, Benjamin Tolchin, Tadeu A Fantaneanu, Andres Fernandez, Lawrence J Hirsch, Suzette LaRoche, Critical Care EEG Monitoring Research Consortium. Association of Periodic and Rhythmic Electroencephalographic Patterns With Seizures in Critically Ill Patients. JAMA Neurol. 2017 Feb 1;74(2):181-188.doi: 10.1001/jamaneurol.2016.49

[24] Aaron F Struck, Berk Ustun , Andres Rodriguez Ruiz, Jong Woo Lee, Suzette M LaRoche, Lawrence J Hirsch, Emily J Gilmore, Jan Vlachy, Hiba Arif Haider, Cynthia Rudin, M Brandon Westover. Association of an Electroencephalography-Based Risk Score With Seizure Probability in Hospitalized Patients. JAMA Neurol. 2017 Dec 1;74(12):1419-1424. doi: 10.1001/jamaneurol.2017.245

[25] Bauer G, Trinka E. Nonconvulsive status epilepticus and coma. Epilepsia. 2010 Feb;51(2):177-90. doi: 10.1111/j.1528-1167.2009.02297.x. Epub 2009 Sep 10

[26] Brandon Foreman, Jan Claassen, Karine Abou Khaled, Jeffrey Jirsch, Daniel M Alschuler, John Wittman, Ronald G Emerson, Lawrence J Hirsch. Generalized periodic discharges in the critically ill: a case-control study of 200 patients. Neurology. 2012 Nov 6;79(19):1951-60.doi: 10.1212/WNL.0b013e3182735cd7.Epub 2012 Oct

[27] San-Juan OD, Chiappa KH, Costello DJ, ColeAJ. Periodic epileptiform discharges in hypoxic encephalopathy: BiPLEDs and GPEDs as a poor prognosis for survival. Seizure. 2009 Jun;18(5):365-8. doi: 10.1016/j.seizure.2009.01.003. Epub 2009 Feb 4.PMID: 19196524

[28] Ficker DM, Westmoreland BF, Sharbrough FW. Epileptiform abnormalities in **hepatic encephalopathy**. J Clin Neurophysiol. 1997 May;14(3):230-4. doi: 10.1097/00004691-199705000-00008. PMID: 9244163

[29] Lai CW, Gragasin ME. Electroencephalography in herpes simplex encephalitis. J Clin Neurophysiol. 1988 Jan;5(1):87-103. doi: 10.1097/00004691-198801000-00003. PMID: 3150762

[30] B K Madhusudhan. Nonconvulsive status epilepticus and Creutzfeldt-Jakob-like EEG changes in a case of lithium toxicity. Epilepsy Behav Case Rep. 2014 Nov 28;2:203-5.doi: 10.1016/j.ebcr.2014.08.005.eCollection 2014.

[31] A R Towne, E J Waterhouse, J G Boggs, L K Garnett, A J Brown, J R Smith Jr, R J DeLorenzo. Prevalence of nonconvulsive status epilepticus in comatose patients. Neurology. 2000 Jan 25;54(2):340-5.doi: 10.1212/wnl.54.2.340

[32] Muge Yemisci, Gunfer Gurer, Serap Saygi, Abdurrahman Ciger. Generalised periodic epileptiform discharges: clinical features, neuroradiological evaluation and prognosis in 37 adult patients. Seizure. 2003 Oct;12(7):465-72.doi: 10.1016/s1059-1311(02)00351-5.

[33] Sutter, Raoul. Are We Prepared to Detect Subtle and Nonconvulsive Status Epilepticus in Critically Ill Patients? Journal of Clinical Neurophysiology. 33(1):25-31, February 2016.DOI: 10.1097/WNP.0000000000000216

[34] J Jirsch, L J Hirsch. Nonconvulsive seizures: developing a rational approach to the diagnosis and management in the critically ill population. Clin Neurophysiol. 2007 Aug;118(8):1660-70.doi: 10.1016/j.clinph.2006.11.312. Epub 2007 Jun 27.

[35] Foreman B, et al. Generalized periodic discharges and 'triphasic waves': A blinded evaluation of inter-rater agreement and clinical significance.Clin Neurophysiol. 2016. PMID: 26294138

[36] Jadeja N, Zarnegar R, Legatt AD. Clinical outcomes in patients with generalized periodic discharges. Seizure. 2017 Feb;45:114-118. doi: 10.1016/j.seizure.2016.11.025. Epub 2016 Dec 7.

[37] Barry J Ruijter , Michel J A M van Putten, Jeannette Hofmeijer Generalized epileptiform discharges in postanoxic encephalopathy: Quantitative characterization in relation to outcome. Epilepsia. 2015 Nov;56(11):1845-54.doi: 10.1111/epi.13202.Epub 2015 Sep 19.

[38] de la Paz D, Brenner RP .Bilateral independent periodic lateralized epileptiform discharges. Clinical significance.Arch Neurol. 1981 Nov;38(11):713-5.doi: 10.1001/archneur. 1981. 00510110073012. PMID: 7305701

[39] Dericioglu N, Khasiyev F, Arsava EM, Topcuoglu MA. Frontal

Intermittent Rhythmic Delta Activity (FIRDA) in the Neurological Intensive Care: Prevalence, Determinants, and Clinical Significance. Clin EEG Neurosci. 2018 Jul;49(4):272-277. doi: 10.1177/1550059416688108. Epub 2017 Jan 24.

[40] Gaspard N, Manganas L, Rampal N, Petroff OA, Hirsch LJ. Similarity of lateralized rhythmic delta activity to periodic lateralized epileptiform discharges in critically ill patients. JAMA Neurol. 2013 Oct;70(10):1288-95. doi: 10.1001/jamaneurol.2013.3475. PMID: 23921464

Chapter 8

Basic Electroencephalogram and Its Common Clinical Applications in Children

Raafat Hammad Seroor Jadah

Abstract

Electroencephalography (EEG) is a non-invasive neurophysiological study that monitors electrical activity of the brain. EEG is an essential investigational tool to analyze and record electrical impulses of the brain and considered to be the gold standard electrophysiological test which can be used to help diagnose epilepsy. EEG can also be used to diagnose and evaluate other conditions such as sleep disorders, neurometabolic diseases with encephalopathy and neuropsychiatric disorders. It is also an essential ancillary test in other conditions such as brain death assessment. However, it is essential not to entirely rely on EEG for an absolute diagnosis of epilepsy as the main indication of EEG in general and in Pediatric age group in particular is to categorize different types of seizure and epilepsy syndromes for further evaluation and management.

Keywords: electroencephalography, epilepsy, neuropsychiatric, ancillary, electrophysiological, pediatric, encephalopathy

1. Introduction

EEG is a common, non-invasive and essential electrophysiological technique used to evaluate and study the brain function. EEG measures and investigates the cerebral electrical impulses by direct application of electrodes to the patient's scalp. EEG is considered the main neurophysiological study used in Pediatric population especially in children with epilepsy [1, 2] and remains the primary test used to study and assess other clinical conditions such as parasomnia and encephalopathy associated with neurometabolic disorders and post traumatic brain injury [3]. EEG study has been used in the evaluation and assessment of organic brain pathology in patients presented with psychiatric and behavioral disorders and has been also an essential tool to confirm absence of cerebral electrical activity in patients with brain death [4, 5]. Epilepsy diagnosis is primarily made based on the clinical history of the patient and hence it is necessary not to rely completely on the EEG study to confirm the diagnosis of epilepsy [6], however EEG is the major neurophysiological test used in the classification and evaluation of seizures and epilepsy syndrome in Pediatric patients [7].

1.1 History

In 1875, Richard Caton an English physician reported a spontaneous electrical variation from exposed cortical brain hemispheres of rabbits and monkeys [8]. Early in the twentieth century, specifically in 1912, Vladimir Vladimirovich Pravdich-Neminsky a Russian Physiologist reported the first electrical brain impulse and evoked response in animals (dog) [8]. However, in 1924 German Neurologist and Psychiatrist Hans Berger recorded the first human EEG in a graph paper which later named an electroencephalogram (EEG) device. Berger subsequently characterized different rhythmic nature and wave patterns of the brain activity based on the different physiological state of the subjects (**Figure 1**) [8]. The initial description of clinical encephalography was first reported by an American neurologist Frederic Andrews Gibbs in 1935 who initially documented the classical interictal spikes associated with epilepsy and first to demonstrate the typical 3 per second spike and wave discharges associated with absence epilepsy. He also described EEG pattern during impaired consciousness level (**Figure 2**) [8–10].

Figure 1.
Hans Berger, German neurologist and psychiatrist (1873–1941) [11].

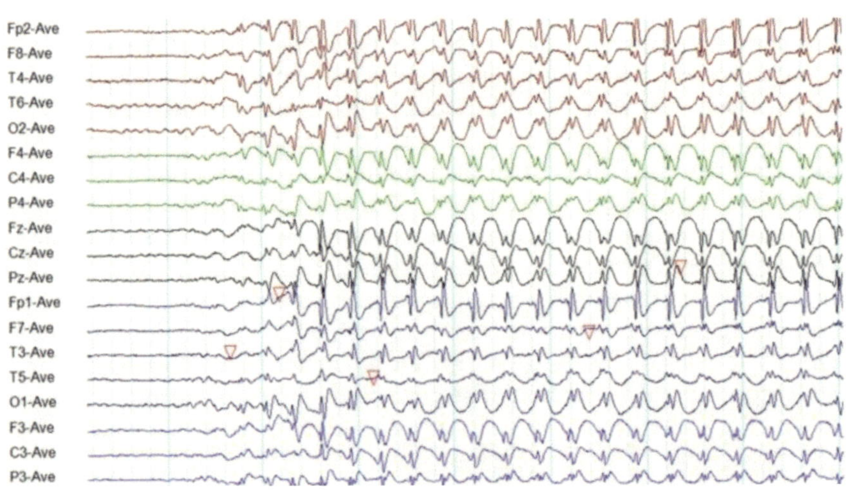

Figure 2.
The classical 3 per second spike and wave discharges was first described by Frederic Gibbs [10].

2. Analysis and understanding the complex brain network

The human brain consists of a complete and comprehensive network map of neuronal connections called human connectome. The normal maturation of these interconnected neurons associated with normal development of high cortical functions and motor skill consolidation. The failure of this network maturation can lead to some serious neurodevelopmental disabilities [12].

The connectivity of this complex brain network can be classified into three types: structural connectivity, functional connectivity and effective connectivity [13]. Structural connectivity can be further subdivided into two types. First the anatomical connections that links a bundle of neural elements and second is the interregional fibers linking cortical to subcortical gray matter areas [13].

Functional connectivity is obtained from time series analysis and reflects the statistical dependence within neural units. This time sense date can be defined by different methods which include EEG, Functional magnetic resonance imaging (fMRI) and magnetoencephalography (MEG) [13].

Effective connectivity (EC) defines the casual effects that one neural system exerts over another. EC cannot be assessed directly so several techniques have been used to study the EC. The Dynamic Casual Modeling (DCM) is the main method for evaluating EC by analyzing data from neuroimaging studies such as Functional Magnetic Resonance Imaging (fMRI) [14].

3. Preparing pediatric patients for EEG study

Performing electroencephalography (EEG) in children can be quite challenging as most of these children are not cooperative during this study due to the great fear and restlessness during the EEG procedure. It is vitally important to prepare a Pediatric patient for EEG study in order to have better interpretation of the EEG results. The application of psychological technique prior to the study and the availability of the parents during the procedure can be helpful to conduct the study smoothly and minimize the need for premedication drugs. However, the behavioral and psychological techniques are not always successful in a small proportion of children. Different premedication protocols have been proposed in order to alleviate the great distress and anxiety during the study. The ideal pharmacological agents for such procedure should have a minor impact on the EEG tracing with fast onset and few side effects. Benzodiazepine is the most common premedication agent used with Midazolam being the most popular drug to induce sedation for EEG study in children [15].

Chloralhydrate is another medication which has been used to induce sedation in the Pediatric population during different neurological studies including EEG. Chloralhydrate is a safe, cheap hypnotic non-opiate drug with no major side effects with the exception of vomiting in few cases. Chloralhydrate has been also shown to be effective and more time saving during EEG procedure [16].

4. Technical aspects of electroencephalography

Electroencephalography EEG, since it's first introduction early in the 20th century, has been an essential and the most common neurophysiological device to monitor and study the electrical and functional activity of the brain [17]. EEG is a commonly used non-invasive tool to track and record the electrical field potentials captured by electrodes placed on the patient scalp. These electric field potentials created by dipoles as a result of excitation of the epical dendritic postsynaptic potential at the cortical pyramidal cells [18, 19]. The measurement and assessment of the electric field potentials can be made by attaching conductive electrodes to the human scalp. At the present time the wet electrodes are the gold standard used for EEG study [19]. A conductive paste or gel need to be used during the application of wet electrodes to minimize electrode-skin impedance in order to achieve good conductivity of the electrical impulse. The typical value of skin impedance should be kept between 5 and 20 KΩ. This skin impedance should be continuously monitored during the EEG study to ensure proper and high-quality conductivity between the skin and the EEG electrode. Performing an EEG study is a time consuming process which require an expert EEG technician or neurophysiologist in order to obtain good quality EEG results for proper interpretation and reporting as the reading and analyzing EEG data is a hard task and must be interpreted by expert neurophysiologists. The location site and description of the scalp electrodes is well recognized by the international 10–20 system (**Figure 3**) [19, 20].

During the first EEG only 20–50% of patients with seizure disorder show interictal epileptiform discharges (IED) so the yield of the EEG study can be enhanced by many activation methods in order to capture the interictal epileptiform discharges which help confirming the diagnosis of epilepsy and seizure disorder [22]. The common activation procedure used in EEG laboratories includes Hyperventilation, intermittent photic stimulation (IPS), sleep and sleep deprived techniques. Hyperventilation (HV) is considered to be the first and oldest activation method used to trigger the interictal epileptiform discharges (IED) especially the one associated with absence epilepsy. HV is more effective in Pediatric population than in adult. A proper effective HV should be carried out for full 3 minutes with continuous recording and monitoring for one-minute post hyperventilation. HV is more efficient in diagnosing generalized seizures than focal epilepsy. The mechanism of HV to trigger interictal epileptiform discharges can be explained by hypocapnia induction which also manifest as background slowing or focal slowing in the EEG [22]. HV is a major provocation technique used to trigger the typical 3-Hz spike-and-wave discharge (SWD) which is characteristic for absence epilepsy as more than 90% of patients who have absence epilepsy show SWD during HV. The non-specific thalamic projection system (NSTPS) which is a part of the thalamocortical networks triggered by respiratory alkalosis and considered to be the major induction of SWD associated with absence epilepsy during the process of HV [23]. HV is an efficient and safe activation method for epilepsy and seizure disorder provocation however there are certain contraindication to perform HV during EEG study which includes patients with cardiopulmonary disease, sickle

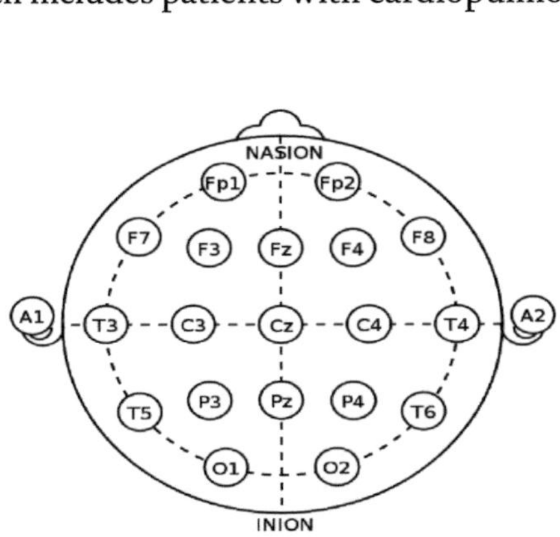

Figure 3.
The international 10-20 system [21].

cell anemia, Moy-Moya disease, subarachnoid and intracerebral bleeding and severe carotid stenosis [22].

A standard activation procedure used during the routine EEG study is the intermittent photic stimulation. This procedure done in a dimmed light room and application of different light frequencies between 1 and 30 Hz for 5 to 10 seconds during eye closure. The flashing light device should be kept 30 cm form the patient eyes. The response to intermittent photic stimulation (IPS) can be seen as an evoked potentials at frequencies less than 5 Hz seen posteriorly or drive response at the occipital regions or in the form of photoparoxysmal response (PPR) which was previously named photoconvulsive response. The most common types seizure disorder seen with IPS are absence epilepsy, myoclonic and tonic–clonic seizures [22].

Among the activation techniques used during routine EEG study is the sleep and sleep deprived approach which produce the maximum yield of interictal epileptiform discharges (IED) as compared to the hyperventilation and intermittent photic stimulation procedure. The young age patients tend to have better yield of IED with each activation technique than older patients [24].

5. Brain computer interface

A Brain Computer Interface (BCI) which also named as Brain –machine Interface (BMI) is a computer- build network system that allow direct communication between cerebral brain activity and external recordable machine without using human muscles or peripheral nervous system. BCI utilize and analyzes the brain signals to collect information and send them to output system. BCI network consists of five phases: Signal Acquisition, Signal Magnification, Feature Extraction, Categorization and Control Interface. BCI assesses and analyze brain activity through mainly electrophysiological and hemodynamic studies. The electrophysiological study consists mainly of EEG, electrocorticography and magnetoencephalography. The hemodynamic study measures glucose uptake by an active neurons and this can be evaluated by procedures like functional magnetic resonance and infrared spectroscopy. BCIs commonly used EEG to gain details from brain activity. The design of BCI is complex due to restricted resolution and data reliability detected by the brain [25, 26].

6. Different types of EEG study

The American Clinical Neurophysiology Society suggest at least 20 minutes' time duration for routine outpatient study. However, the International League against Epilepsy suggests a minimum 30 minute for routine EEG recording. Currently most routine EEG studies are done with an average time between 20 and 30 minutes. The abnormal epileptiform discharges found in 29–55% in patients with epilepsy on their first routine EEG study. Ambulatory prolonged EEG study is considered to be helpful diagnostic technique to capture interictal epileptiform discharges (IEDs) in epilepsy patients whom their first routine EEG studies reported normal. Prolonged ambulatory EEG study is considered to be superior to routine EEG in identifying IEDs specially during the natural sleep state. This procedure is also helpful to differentiate epileptic from non-epileptic psychogenic events. The duration of the ambulatory EEG study usually between 24 to 96 hours [27, 28].

Epilepsy Monitoring Unit (EMU) is an important and crucial part of the neurophysiological work up for the diagnosis and classification of epilepsy and evaluation of psychogenic non-epileptic seizures (PNES). EMU is also essential for patients with intractable epilepsy resistant to antiepileptic medications and for evaluating candidates for possible epilepsy surgery [29].

EMU is strongly recommended for children with unclear history of paroxysmal episodes in order to differentiate between epileptic and non-epileptic events as this can be quite challenging in pediatric population. EMU is also important in evaluating different types of epilepsy syndromes in children. One of the vital advantages of the video-EEG telemetry is the monitoring and recording the ictal events especially in patients with partial epilepsy. EMU is a highly selective study should be done for carefully selected patients as this is an expensive and time consuming procedure [30]. The process of monitoring and recording video-EEG telemetry can range from 24 hours to 7 days. In some situation antiepileptic drugs need to be tapered in order to induce seizure activity for better evaluation of seizure semiology and localization of the epileptogenic zone [31].

The Amplitude-Integrated EEG (aEEG) is another continuous electrophysiological modality used in both term and preterm newborns in the neonatal intensive care units (NICUs). Since it's first introduction late in 1980s. the aEEG considered to be the gold standard to monitor and assess neonatal brain background activity, diagnose and manage newborn seizure disorders and help in selecting newborns who might be benefit from cooling therapy. The aEEG also plays a major role in predicting the neurodevelopmental outcomes for term and preterm newborn babies. The application and recording of the aEEG is done by using two or four scalp electrodes applied to C3, P3, C4 and P4 positions of the newborn head according to the international 10–20 system. aEEG is a safe procedure which has a major limitation as it covers only small area of the head surface and hence focal epileptiform activity cannot be monitored during the aEEG recording [32–34].

EEG is considered to be the commonest procedure used for intraoperative neurophysiological and cerebral perfusion monitoring [35]. EEG is also considered to be the gold standard modality for evaluating patients for possible epilepsy surgery to localize and define different epileptogenic foci. EEG also plays an important role in understanding the nature and pathophysiology of epilepsy and presurgical evaluation of functional cortical mapping. However, routine EEG monitoring might not be always sufficient to evaluate certain types of epilepsy such as non-lesional temporal lobe epilepsy which necessitate the need of more interventional procedure such as the invasive electroencephalography (iEEG) [36, 37].

7. Basic EEG interpretation

A proper and detailed history taking is more reliable and important in diagnosing epilepsy and seizure disorders than EEG study. A solid and classical history of seizure even with the presence of normal EEG finding make the diagnosis of epilepsy is more likely as the sensitivity of single routine EEG study is only about 50% in diagnosing seizure disorders [38].

A single EEG study provides extensive data for interpretation. The main initial description of the EEG recording includes the amplitude, frequency and wave morphology. Hans Berger described two characteristic EEG wave frequencies during awake state: The alpha rhythm (8–12 Hz) which is more prominent in the arousable stage with eye closure and beta rhythm (13–30 Hz) commonly seen with mind focus state. In most people eye closure will result in frequency transfer from beta to alpha rhythms. Subsequent wave frequencies were identified the theta rhythm (4–7 Hz) and the delta rhythm (0.5–3 Hz) which are predominant during sleep in adults, and the gamma rhythm (> 30 Hz) which is associated with memory, information processing and cognitive skill (**Figure 4**) [39].

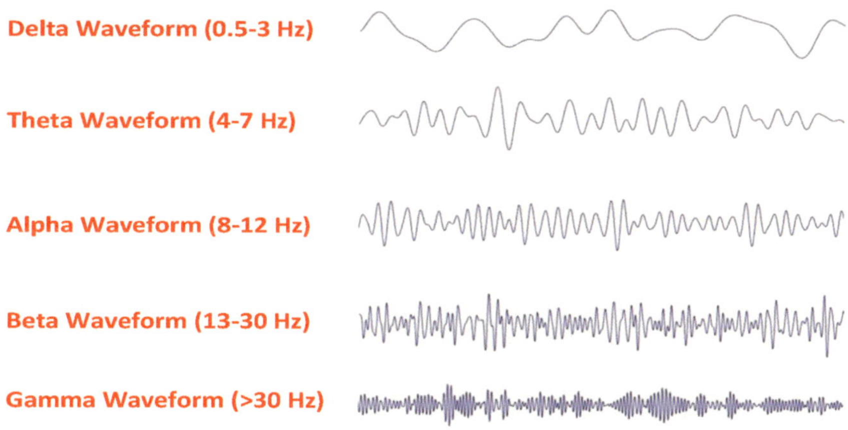

Figure 4.
Different EEG waveforms [39].

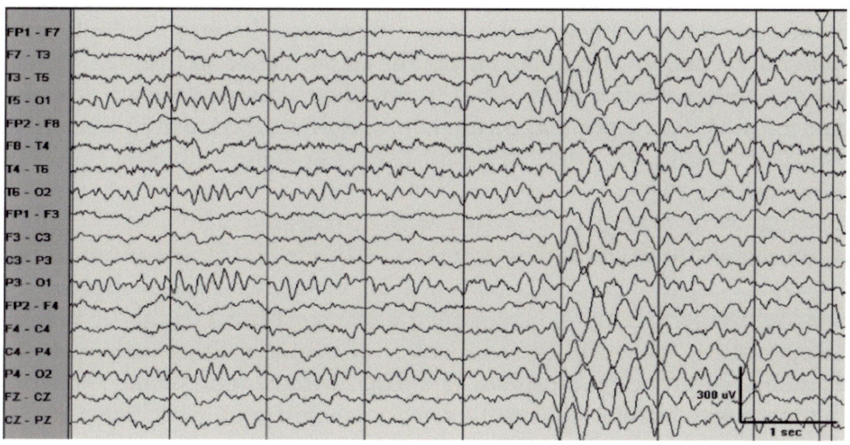

Figure 5.
Hypnagogic hypersynchrony. A normal EEG variant [42].

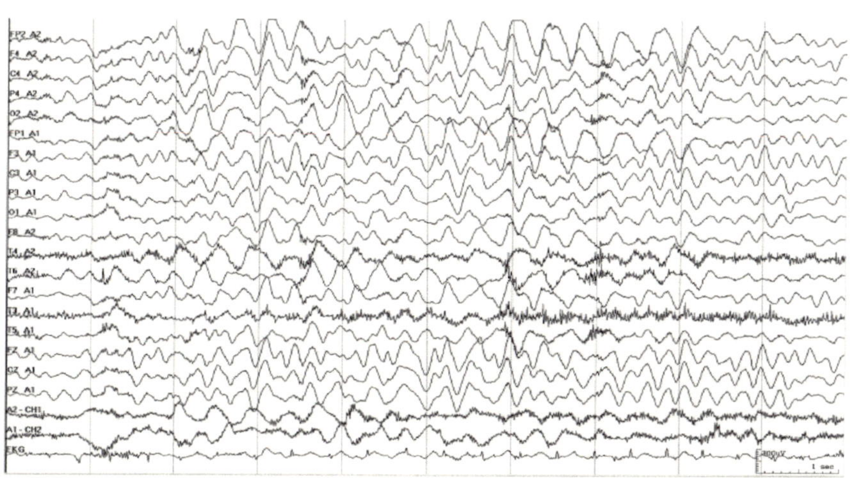

Figure 6.
High amplitude rhythmic slowing with hyperventilation [43].

It is vitally important to ensure proper education and gain enough experience to read and interpret EEG recording in order to avoid misdiagnosis of epilepsy and to provide better care to the patient. It is also essential to appreciate the common benign variations of normal EEG study [40, 41].

A common normal patterns seen in EEG study which can be falsely interpreted as abnormal epileptiform discharges include multifocal sharp waves and spikes, generalized slowing with hyperventilation, hypnagogic hypersynchrony and most commonly is the background alteration at the temporal area (**Figures 5** and **6**) [38].

Over interpretation of normal EEG tracing is the main factor for misdiagnosis of epilepsy and seizure disorders. Improper neurophysiological training and inadequate experience is the major reason for over interpretation of normal EEG study. Conservative EEG interpretation and avoiding biased history are strongly recommended by all epileptologists [38].

8. Common clinical applications of EEG in children

Although the diagnosis of epilepsy is primarily made by clinical history, EEG remains an essential investigational tool to differentiate between epileptic and non-epileptic events, it's also important in the classification of different types of epilepsy and epilepsy syndromes [44–46]. Frequent classical epileptiform abnormalities seen in Pediatric population are hypsarrhythmia associated with infantile spasm, 3 Hz spike and wave discharges in absence epilepsy and burst suppression (**Figures 7** and **8**) [46].

According to the American Academy of Neurology and Child Neurology Society, EEG is recommended in children presented with their first attack of unprovoked seizure [49]. EEG is indicated in children with atypical febrile convulsion or prolonged febrile seizure and it is an essential investigational study in patients with newly diagnosed epilepsy and in classification of common childhood epilepsy syndromes such as centrotemporal spikes associated with benign rolandic epilepsy and Panayiotopoulos syndrome (idiopathic childhood epilepsy). EEG is also important in recording continuous spike-waves during slow-wave sleep (CSWS) in epileptic encephalopathies (**Figures 9** and **10**) [50].

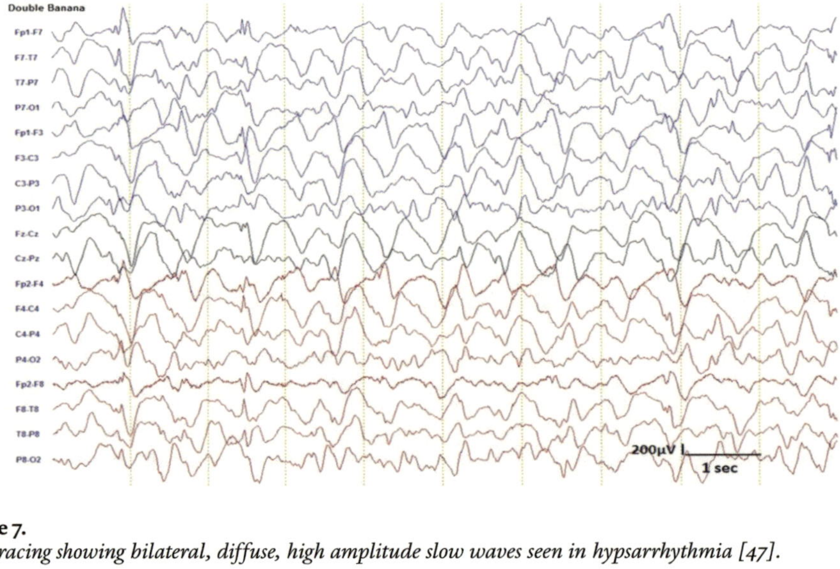

Figure 7.
EEG tracing showing bilateral, diffuse, high amplitude slow waves seen in hypsarrhythmia [47].

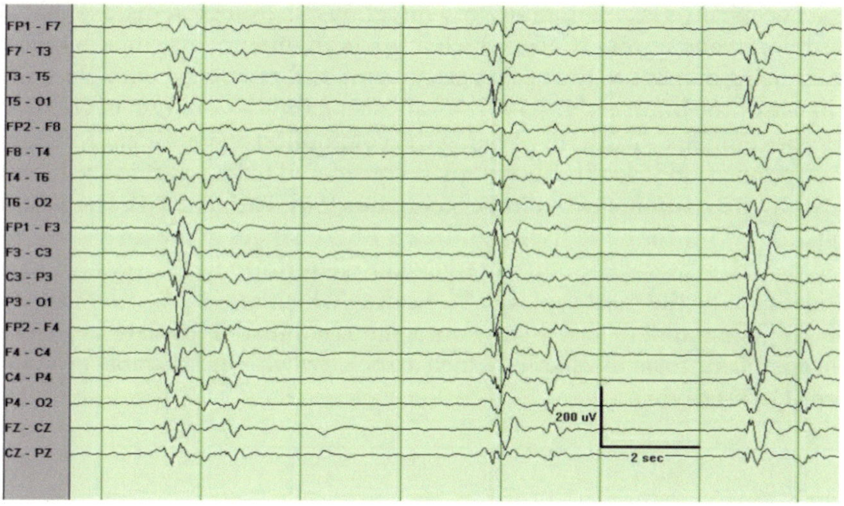

Figure 8.
EEG with burst suppression [48].

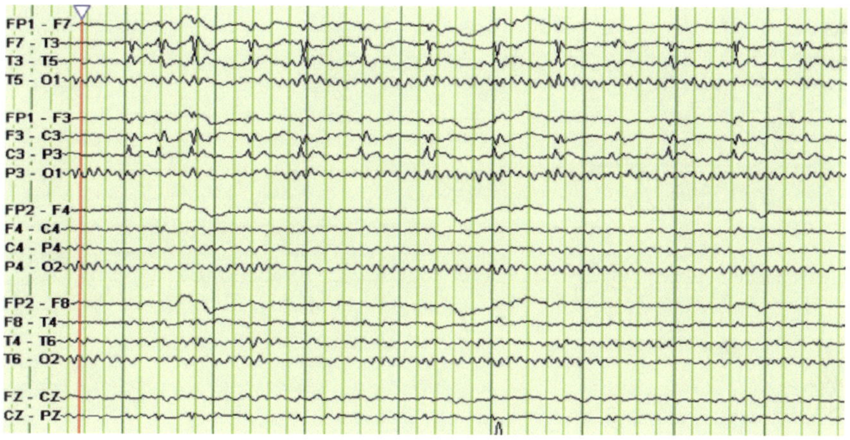

Figure 9.
EEG showing left centrotemporal epileptiform spike and wave discharges in patients with benign rolandic epilepsy [51].

Pediatric patients diagnosed with autistic spectrum disorder (ASD) with positive history of epilepsy and abnormal findings in the neurological examination, EEG study is indicated as a part of their screening tests. EEG is also recommended in monitoring antiepileptic medication in patients with confirmed diagnosis of epilepsy [50].

The background EEG monitoring has been also used in children with traumatic brain injury which is helpful in evaluating prognosis in these patients [50, 53]. EEG study with poor reactivity, prolonged discontinuity and burst suppression associated with poor prognosis, whereas EEG with good reactivity and normal sleep rhythm favor a good prognosis [53]. A prolonged EEG recording is also essential in children admitted to the PICU (Pediatric Intensive Care Unit) with suspected non-convulsive seizures (NCS). It has been also important in monitoring Pediatric patients underwent surgery for congenital cardiac anomalies as they are at risk to have seizure post-surgery [53].

Viral encephalitis is an inflammatory infectious neurological disease that affects the central nervous system (CNS) which triggers an immune response by the viral antigen causing damage to the brain parenchyma and associated with electrical disturbance of the brain activity [54]. Viral encephalitis is common in children with Herpes Simplex Virus (HSV) being the commonest agent for encephalitis in Pediatric population [54, 55]. EEG study is considered to be a part in the investigational work up in patients with viral encephalitis [56]. Patients with Herpes Simplex Encephalitis (HSE) found to have significant EEG changes in the early stage of the diagnosis. Unilateral Periodic Lateralized Epileptiform Discharges (PLED) is considered to be the most typical EEG finding which correlate with diagnosis of HSE [57] and found to have a good outcome as compared to bilateral periodic lateralized epileptiform discharges which associated with unfavorable prognosis (**Figure 11**) [58].

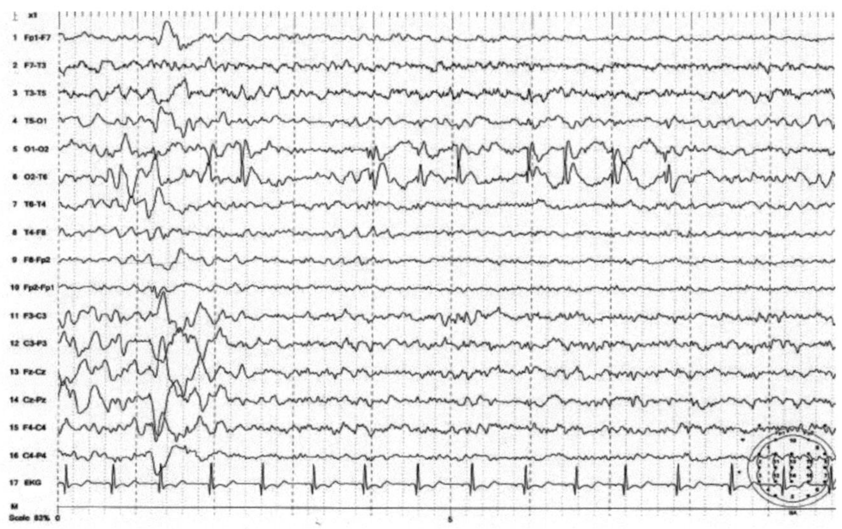

Figure 10.
Occipital spike and wave discharges seen in panayiotopoulos syndrome [52].

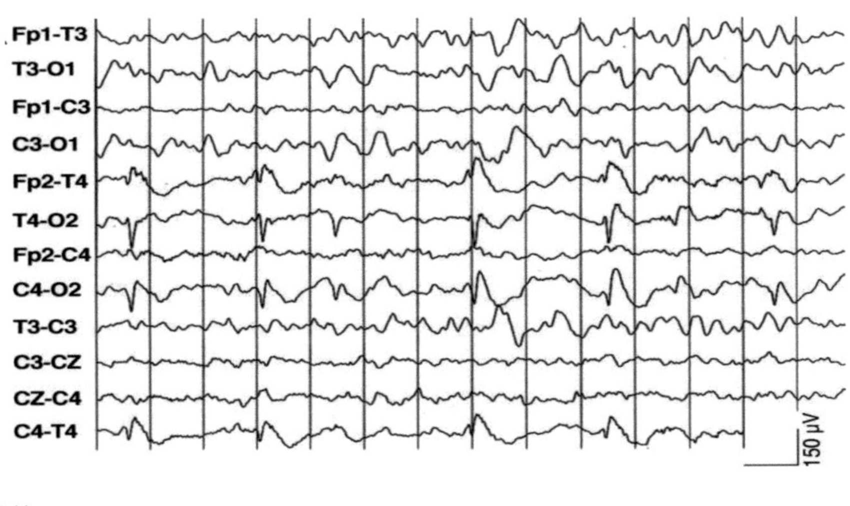

Figure 11.
Periodic lateralized epileptiform discharges (PLED) over the right central-temporal head regions seen in HSE [59].

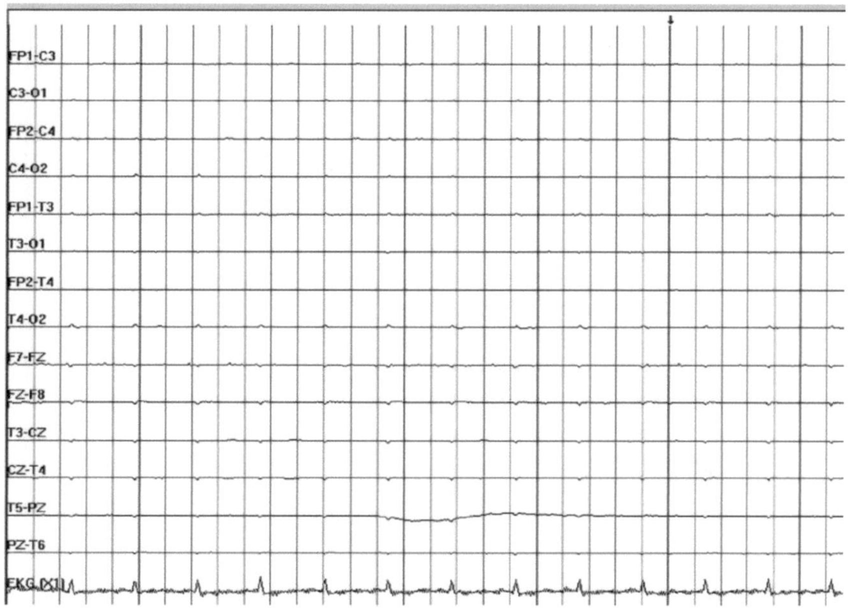

Figure 12.
Isoelectric EEG. No cerebral brain activity with sensitivity over 2 μV/mm [62].

EEG study can be also used as an ancillary test to support the diagnosis of brain death. Although positive diagnosis of brain death can be made by two separate settings of clinical examination, The American Neurological Association strongly suggest the use of EEG study to confirm the diagnosis of brain death. Hypothermia and hypotension should be avoided when applying EEG for brain death assessment [60]. Isoelectric encephalogram is confirmed when 30 minutes of EEG recording reveals complete absence of cerebral activity with sensitivity over 2 μV/mm in the absence of electrolyte disturbance and sedative medications (**Figure 12**) [61].

9. Conclusion

EEG is considered to be save non-invasive procedure since its first application early in the 20th century. This procedure is done by trained EEG technicians and it should be interpreted by Neurologist or expert Neurophysiologists in order to obtain a high quality report to reach a proper diagnosis and provide optimal management to the patient.

Performing EEG study in children can be a difficult task because of the great fear and anxiety in this age group patients, so its vitally important to properly prepare these patients to minimize EEG artifacts for better interpretation of the EEG report.

EEG is an essential neurophysiological study especially in Pediatric patients to differentiate epileptic form and non-epileptic events as the differential diagnosis for paroxysmal episodes in children is wide and varies according to certain age group.

Although the diagnosis of brain death is primarily made on clinical basis, EEG remains an important ancillary test for diagnostic confirmation of brain death.

Conflict of interest

The author declares no conflict of interest.

Funding

None.

Abbreviations

EEG	electroencephalography
fMRI	functional magnetic resonance imaging
MEG	magnetoencephalography
IED	interictal epileptiform discharges
IPS	intermittent photic stimulation
BCI	brain computer interface
EMU	epilepsy monitoring unit
aEEG	amplitude-integrated EEG
ASD	autistic Spectrum disorder

Author details

Raafat Hammad Seroor Jadah
Department of Paediatric, Bahrain Defence Force Hospital, Royal Medical Services, Kingdom of Bahrain

*Address all correspondence to: nader212@hotmail.com

References

[1] Gregory A Light, Lisa E Williams, Falk Minow, Joyce Sprock, Anthony Rissling, Richard Sharp, Neal R Swerdlow, David L Braff. Electroencephalography (EEG) and event-related potentials (ERPs) with human particpants. Curr protoc Nuerosci.2010 Jul; Chapter 6: unit6.25.1-24. DOI: 10.1002/0471142301.ns0625s52.

[2] Mohammed M S Jan Assessment of the utility of Paediatric electroencephalography. Seizure.2002 Mar; 11(2):99-103. DOI: 10.1053/seiz.2002.0621.

[3] C D Binnie, P F Prior Electroencephalography J Neurol Neurosurg Psychiatry. 1994 Nov; 57(11): 1308-19.DOI: 10.1136/jnnp.57.11.1308.

[4] S S O'Sullivan, G M Mullins, E M Cassidy, B McNamara The role of the standard EEG in clinical psychiatry. Hum psychopharmacol. 2006 Jun; 21(4): 265-71. DOI: 10.1002/hup.767.

[5] Daniel Kroeger, Bogdan Florea, Florin Amzica. Human brain activity patterns beyond the isoelectric line of extreme deep coma. PLoS One. 2013 sep 18; 8(9): e75257.13 pages (1-13).

[6] Ushtar Amin, Selim R Benbadis. The Role of EEG in the Erroneous Diagnosis of Epilepsy. J Clin Neurophysiol. 2019 Jul; 36(4): 294-297. DOI: 10.1097/WNP.0000000000000572.

[7] M Sundaram, R M Sadler, G B Young, N Pillay. EEG in epilepsy: Current perspectives. Can J Neurol Sci. 1999 Nov; 26(4): 255-62. DOI: 10.1017/s0317167100000342.

[8] Mahmoud Al-Kadi, Mamun Bin Ibne Reaz, Mohd Alauddin Mohd Ali. Evolution of Electroencephalogram Signal Analysis Techniques During Anesthesia Sensors (Basel). 2013 May; 13 (5): 6605-6635. DOI: 10.3390/s130506605

[9] Prasad Vannemreddy, James L Stone, Konstantin V slavin. Frederic Gibbs and his contributions to epilepsy surgery and electroencephalography. Neurosurgery.2012 Mar; 70 (3): 774-82. DOI: 10.1227/NEU.0b013e3182351699.

[10] Hal Blumenfeld. Consciousness and epilepsy: why are patients with absence seizures absent? Prog Brain Res. 2005; 150:271-286 DOI: 10.1016/S0079-6123(05) 50020-7

[11] James L Stone, John R HughesEarly history of electroencephalography and Establishment of the American Clinical Neurophysiology Society. J Clin Neurophysiol. 2013 Feb; 30(1): 28-44 DOI: 10.1097/WNP.0b013e31827edb 2d.

[12] Stuart Oldham, Alex Fornito. The development of brain network hubs Dev Cogn Neurosci. 2019 Apr; Volume 36:100607 Number of pages 60 (1-60) DOI : 10.1016/j. dcn. 2018.12.005.

[13] Olaf Sporns. Structure and function of complex brain networks Dialogues Clin Neurosci. 2013 sep; 15(3): 247-262. Retrieved from https://www.ncbi.nlm.nih.gov/pmc/articles/PMC3811098

[14] Peter Zeidman, Amirhossein Jafarian, Nadege Corbin, Mohmed L. Seghier, Adeel Razi, Cathy J. Price, Karl J. Friston. A guide to group effective connectivity analysis, part 1: first level analysis with DCM for fMRI. Neuroimage. 2019 oct 15; 200: 174-190 DOI: 10.1016/j.neuroimage.2019.06.031.

[15] Mahmoud Reza Ashrafi, Hossein Mohebbi, Mahmoud Mohammadi, Elham Azizi, Gholam Reza Zamani, Alireza Tavasoli, Reza Shervin Bady, Firozeh Hosseini. Clonidine versus

Chloral Hydrate for Recording sleep EEG in children Iran J Child Neurol.2020 Winter; 14 (1): 85-92 Retrieved from https://pubmed.ncbi.nlm.nih.gov/32021632.

[16] Razieh Fallah , Sharam Jalili, Motahhareh Golestan, Sedighah Akavan Karbasi, Mohammad-Hosein Jarahzadeh. Efficacy of Chloral Hydrate and Promethazine for Sedation during Electroencephalography in Childern; a Randomised Clinical Trial. Iran J pediatr.2013 Feb ;23(1) Retrieved from http://www.ncbi.nlm.nih.gov/pmc/articles/PMC3574988

[17] Hector Rieiro, Carolina Diaz-Piedra, Jose Miguel Morales, Andres Catena, Samuel Romero, Joaquin Roca-Gonzalez, Luis J Fuentes, Leandro L Di Stasi. Validation of Electroencephalographic Recordings Obtained with a Consumer-Grade, Single Dry Electrode, Low-Cost Device: A Comperative Study. Sensors (Basel).2019 Jun 23; 19(12): 2808 18 pages (1-18) DOI: 10.3390/s19122808.

[18] Alexander J Casson Wearable EEG and beyond. Biomed Eng Lett. 2019 Jan 4; 9(1): 53-71.DOI: 10.1007/s13534-018-00093-6.

[19] M.A. Lopez –Gordo, D. Sanchez-Morillo, F.Pelayo Valle. Dry EEG Electrodes Sensors (Basel). 2014 Jul; 14(7): 12847-12870. DOI:10.3390/s140712847.

[20] Mark O'Sullivan, Andriy Temko, Andrea Bocchino, Conor O'Mahony, Geraldine Boylan, Emanuel Popovici. Analysis of a Lost-Cost EEG Monitoring System and Dry Electrodes toward Clinical use in the Neonatal ICU. Sensors (Basel) .2019 Jun; 19(11): 2637 16 pages (1-16). DOI: 10.3390/s19112637.

[21] C Kabdebon, F Leroy, H Simmonet, M Perrot, J Dubois, G Dehaene-Lambertz. Anatomical Correlations of the International 10-20 sensor placement system in infants. Neuroimage.2014 Oct 1; 99:342-56. DOI: 10.1016/j.neuroimage.2014.05.046.

[22] Oscar E Mendez, Richard P Brenner. Increasing the yield of EEG. Clin Neurophysiol. 2006 Aug; 23(4): 282-93. DOI: 10.1097/01.wnp.0000228514.40227.12.

[23] Kathryn A. Salvati, Mark P. Beenhakker. Out of thin air: Hyperventilation-triggered seizures. Brain Res. 2019 Jan 15; 1703:41-52. DOI: 10.1016/j.brainres.2017.12.037

[24] Elisa Baldin, W A Hauser, Jeffrey R Buehhalter, Dale C Hesdorffer, Ruth Ottman. Utility of EEG Activation Procedures in Epilepsy: A Population-Based Study. J Clin Neurophysiol.2017 Nov; 34 (6): 512-519. DOI: 10.1097/WNP.0000000000000371.

[25] Jerry J. Shih, Dean J. Krusienski, Jonathan R.Wolpaw.Brain-Computer Interfaces in Medicine. Mayo Clin Proc. 2012 Mar; 87(3):268-279. DOI: 10.1016/j.mayocp.2011.12.008

[26] Luis Fernando Nicolas-Alonso, Jaime Gomez-Gil. Brain – Computer Interfaces, a Review. DOI: 10.3390/s120201211

[27] David B. Burkholder, Jeffrey W. Britton, Vijayalakshmi Rajasekaran, Rachel R. Fabris, Perumpillichira J. Cherian, Kristen M. Kelly-Williams, Elson L. So, Katherine C. Nickels, Lily C. Wong-Kisiel, Terrence D. Lagerlund. Gregory D. Cascino, Gregory A. Worrell, Elaine C. Wirrell. Routine Vs extended outpatient EEG for the detection of interictal epileptiform discharges. Neurology. 2016 Apr 19; 86(16): 1524-1530. DOI: 10.1212/WNL.0000000000002592.

[28] Udaya Seneviratne, Wendyl Jude D'Souza. Ambulatory EEG. Handb Clin Neurol. 2019; 160: 161-170. DOI: 10.1016/B978-0-444-64032-1.00010-2.

[29] Brian D Moseley, Sandra Dewar, Zulfi Haneef, Dawn Eliashiv, John M Stern. Reasons for prolonged length of stay in the epilepsy monitoring unit. Epilepsy Res. 2016 Nov; 127:175-178. DOI: 10.1016/j.eplepsyres.2016.08.030.

[30] O Bennett- Back, S Uliel-Siboni, U Kramer. The yield of video- EEG telemetry evaluation for non-surgical candidate children. Eur J Paediatr Neurol. 2016 Nov; 20(6): 848-854. DOI: 10.1016/j.ejpn.2016.05.017.

[31] Jayanti Mani Video electroencephalogram telemetry in temporal lobe epilepsy Ann Indian Acad Neurol. 2014 Mar; 17 (Suppl 1): S45-S49. DOI: 10.4103/0972-2327.128653.

[32] Hannah C. Glass, Courtney J. Wusthoff, Renee A. Shellhaas. Amplitude Integrated EEG: The Child Neurologist's Perspective J Child Neurol. 2013 Oct; 28(10): 1342-1350. DOI: 10.1177/0883073813488663

[33] Nora Bruns, Susanne Blumenthal, Irmgard Meyer, Susanne Klose-Verschuur, Ursula Felderhoff-Muser, Hanna Muller. Application of an Amplitude –Inegrated EEG Monitor (Cerebral Function Monitor) to Neonates J Vis Exp.2017 Sept 6; (127): 55985 9 Pages (1-9). DOI: 10.3791/55985

[34] Maliheh Kadivar, Elahe Movahedi Moghadem, Reza Shervin Badv, Raziye Sangsari, Maryam Saeedy. A Comparison of Conventional Electroencephalography with Amplitude Integrated EEG in Detection of Neonatal Seizures. Med Devices (Auckl) . 2019 Dec 10; 12: 489-496. DOI: 10.2147/MDER.S214662.

[35] Michael R Isley, Harvey L Edmonds Jr, Mark Stecker. Guidelines for intraoperative neuromonitoring using raw (analog for digital waveforms) and quantitative electroencephalography: a position statement by the American Society of Neurophysiological Monitoring.J Clin Monit Comput. 2009 Dec; 23(6): 369-90. DOI: 10.1007/s10877-009-9191-y.

[36] Aashit K. Shah, Sandeep Mittal. Invasive electroencephalography monitoring: Indications and presurgical planning. An Indian Acad Neurol. 2014 Mar; 17(Suppl 1): S89-S94 DOI: 10.4103/0972-2327.128668.

[37] Edward H. Bertram Electrophysiology in epilepsy surgery: Roles and limitations. Ann Indian Acad Neurol. 2014 Mar; 17 (Suppl 1): S40-S44 DOI: 10.4103/0972-2327.128649.

[38] Selim R. Benbadis, Kaiwen Lin. Errors in EEG interpreation and misdiagnosis of epilepsy. Which EEG patterns are overread? Eur Neurol. 2008;59 (5): 267-71 DOI: 10.1159/000115641.

[39] Yi Sun, Changwei Wei, Victoria Cui, Meihong Xiu, Anshi Wu. Electroencephalography: Clinical applications during the perioperative period. Front Med (Lausanne). 2020 June; Volume 7, article 251 10 pages (1-10). DOI: 10.3389/fmed.2020.00251.

[40] William O Tatum. Normal "suspicious" EEG Neurology. 2013 Jan 1; 80 (1 Suppl 1): S4-11.DOI: 10. 1212/WNL.0b013e31827974df.

[41] Jose Mari-Acevedo, Kirsten Yelvington, William O Tatum. Normal EEG Variants. Handb Clin Neurol.2019; 160:143-160. DOI: 10.1016/B978-0-444-64032-1.00009-6.

[42] E M Mizrahi. Avoiding the pitfalls of EEG interpretation in childhood epilepsy. Epilepsia. 1996; 37 Suppl 1: S41-51. DOI: 10.1111/j.1528-1157.1996.tb06021.x

[43] Leanna M Lum, Mary B Connolly, Kevin Farrell, Peter K H Wong. Hyperventilation- induced high-amplitude rhythmic slowing

with altered awareness: a video-EEG comparison with absence seizures. Epilepsia.2002 Nov; 43(11): 1372-8. DOI: 10.1046/j.1528-1157.2002.35101.x

[44] Arif Khan, Aravindhan Baheerathan. Electrocephalogram after first unprovoked seizure in children: Routine, unnecessary or case specific. J Pediatr Nuerosci.2013 Jan- Apr; 8(1): 1-4 DOI: 10.4103/1817-1745.111412.

[45] B Kollar, Z Carnicka, P Siarnik, L Krizova, S Sutovsky, P Traubner, K Klobucnikova. The importance of interictal electroencephalography in paroxysmal states. Bratisl Lek Listy. 2014; 115 (3): 168-70. DOI: 10.4149/bll_2014_168.

[46] K S Rana. Rational use of EEG in childhood epilepsy Indian J Pediatr. 2000 Jan; 67 (1 Suppl): S22-31. Retrieved from https://pubmed.ncbi.nlm.nih.gov/11129891

[47] Natia Japaridze, Muthuraman Muthuraman, Friederike Moeller, Rainer Boor, Abdul Rauf Anwar, Gunther Deuschl, Urlich Stephani, Jan Raethjen, Michael Siniatchkin. Neuronal Network in west Syndrome as revealed by source analysis and renormalized partial directed coherence. Brain Topogr.2013 Jan; 26 (1): 157-70. DOI: 10.1007/s10548-012-0245-y.

[48] Stephen A Thompson, Stephen Hantus. Highly Epileptiform Bursts Are Associated with Seizure Recurrence. J Clin Neurophysiol.2016 Feb; 33(1): 66-71. DOI: 10.1097/WNP.0000000000000232.

[49] A B Chelse, Kent Kelley, Joseph R Hageman, Sookyong Koh. Initial evaluation and management of a first seizure in children. Pediatr Ann. 2013 Dec; 42(12): 244-8. DOI: 10.3928/00904481-20131122-08.

[50] Kaminska A, Cheliout-Heraut F, Eisermann M, Touzery de Villepin A, Lamblin MD. EEG in children, in the laboratory or at the patient's bedside. Neurophysiol Clin. 2015 Mar;**45**(1):65-74. DOI: 10.1016/j.neucli.2014.11.008

[51] P Dryzalowski, S Jozwiak, M Franckiewicz, J Strzelecka. Benign epilepsy with centrotemporal spikes-current concepts of diagnosis and treatment. Neurol Neurochir pol.2018 Nov-Dec; 52(6): 677-689. DOI: 10.1016/j.pjnns.2018.08.010.

[52] R Caraballo, R Cersosimo, C Medina, N Fejerman. Panayiotopoulos-type benign childhood occipital epilepsy: a prospective study. Neurology. 2000 Oct 24; 55(8):1096-1100. DOI: 10.1212/wnl.55.8.1096.

[53] Nicholas S. Abend, Kevin E. Chopamn, William B. Gallentine, Joshua Goldstein, Ann E. Hyslop, Tobias Loddenkemper, Kendall B Nash, James J. Riviello Jr, Cecil D. Hahn. Electroencephalographic Monitoring in the Paediatric Intensive Care Unit. Curr Neurol Neurosci Rep. 2013 Mar; 13(3): 330 10 Pages (1-10). DOI: 10.1007/s 11910-012-0330-3.

[54] Yupeng Wu, Meihua Chen, Yu Cui, Xiying He, Jingzhong Niu, Yanbo Zhang, Li Zhou. Viral encephalitis in quantitative EEG. J Integr Nuerosci.2018;17(3-4): 493-501. DOI: 10.3233/JIN-180084.

[55] Kevin Messacar, Marc Fischer, Samuel R Dominguez, Kenneth L Tyler, Mark J Abzug. Encephalitis in US children Infect Dis Clin North Am.2018 Mar; 32(1): 145-162. DOI: 10.1016/j.idc.2017.10.007.

[56] Michael J. Bradshaw, Arun Venkatesan. Herpes Simplex Virus-1 Encephalitis in Adults: Pathophysiology, Diagnosis and Management. Neurotherapeutics.2016 Jul;13(3):493-508. DOI: 10.1007/s13311-016-0433-7.

[57] C W Lai, M E Gragasin. Electroencephalography in herpes

simplex encephalitis. J Clin Neurophysiol. 1988 Jan; 5(1): 87-103. DOI: 10.1097/00004691-198801000-00003.

[58] Young-Soo Kim, Keun-Hwa Jung, Soon-Tae Iee, Bong Su Kang, Jung Sook Yeom, Jangsup Moon, Jung –Won Shin, Sang Kun Lee, Kon Chu. Prognostic value of initial standard EEG and MRI in patients with Herpes Simplex Encephalitis. J. Clin Neurol.2016 Apr; 12(2): 224-229. DOI: 10.3988/jcn.2016.12.2.224.

[59] Alberto M Cappellari, Giacomo Tardini, Alberto R Bona, Massimo Belli, Fabio Triulzi, Emilio F Fossali. Teaching NeuroImages: Infantile herpes simplex encephalitis. Neurology. 2014 Aug 12: 83(7): e85-6. DOI: 10.1212/WNL.0000000000000695.

[60] Natalie Henderson, Mark J McDonald. Ancillary studies in evaluating Paediatric brain death. J Pediatr Intensive Care. 2017 Dec; 6(4): 234-239. DOI: 10.1055/s-0037-1604015.

[61] W Szurhaj, M-D Lamblin, A Kaminska, H Sediri. EEG guidelines in the diagnosis of brain death. Neurophysiol Clin. 2015 Mar; 45(1): 97-104. DOI: 10.1016/j.neucli.2014.11.005.

[62] Matthew A Koenig, Peter W Kaplan. Brain Death. Handb Clin Neurol .2019; 161: 89-102. DOI: 10.1016/B978-0-444-64142-7.00042-4.